Multiple Sensing-Driven Feature Extraction and Defect Detection

Zhifen Zhang · Guangrui Wen · Wenjing Ren

Multiple Sensing-Driven Feature Extraction and Defect Detection

西安交通大学出版社
XI'AN JIAOTONG UNIVERSITY PRESS

Springer

Zhifen Zhang
School of Mechanical and Engineering
Xi'an Jiaotong University
Xi'an, Shaanxi, China

Guangrui Wen
School of Mechanical Engineering
Xi'an Jiaotong University
Xi'an, Shaanxi, China

Wenjing Ren
School of Materials Science
and Engineering
Tianjin University
Tianjin, Shaanxi, China

ISBN 978-981-96-4807-8 ISBN 978-981-96-4808-5 (eBook)
https://doi.org/10.1007/978-981-96-4808-5

Jointly published with Xi'an Jiaotong University Press
The print edition is not for sale in China (Mainland). Customers from China (Mainland) please order the print book from: Xi'an Jiaotong University Press.

This Springer imprint is published by the registered company Springer Nature Singapore Pte Ltd.
The registered company address is: 152 Beach Road, #21-01/04 Gateway East, Singapore 189721, Singapore

Preface

In the context of the “14th Five-Year Plan” for intelligent manufacturing development, the 2035 vision outline, and the “14th Five-Year Plan” for high-quality manufacturing development, intelligent welding technology has become an indispensable basis for reliability in the field of high-end equipment welding manufacturing. The level of intelligence in welding technology not only enhances product quality and consistency but also significantly increases production efficiency. At the core of intelligent welding is the integration of advanced sensors, data analysis, and artificial intelligence technologies, enabling real-time monitoring and adaptive control of the welding process. The emergence of additive manufacturing technologies derived from welding, such as arc additive manufacturing and laser additive manufacturing (3D printing), has injected new vitality into traditional welding research, facilitating a deeper integration between welding and additive manufacturing. Intelligent welding is not merely an enhancement of manufacturing efficiency; it is also a crucial factor in the industry’s transformation. It renders the manufacturing process more flexible and adaptable, reducing material waste and aligning with sustainable development goals. As a result, intelligent welding technology has once again been elevated to an important and irreplaceable position within the manufacturing industry, providing critical technological pathways for improving manufacturing efficiency and reliability, thus contributing to the achievement of high-quality development objectives.

By integrating spectral analysis with advanced machine learning methods, this book presents an innovative framework for the development of an intelligent automatic welding monitoring system. This framework centers on spectral technology, utilizing spectral sensors to capture real-time changes in the welding pool. Concurrently, machine learning algorithms serve as the backbone of the system, enabling rapid detection and identification of welding defects. The system not only provides real-time feedback on welding quality but also adaptively optimizes welding parameters based on monitoring results. This integrated approach, combining spectral technology and machine learning, signifies the future direction of intelligent welding manufacturing.

This book is divided into five chapters. Chapter 1 reviews the related work in the field of spectral sensing defect detection and control in welding processes. A GTAW internal defect monitoring method based on optical spectroscopy is proposed in Chap. 2. Based on this, an optical spectroscopy-based GTAW multi-defect monitoring method is proposed in Chap. 3, which is capable of identifying the coupling between multiple defects simultaneously. Chapter 4 presents a real-time prediction method for laser beam welding weld strength based on optical spectroscopy. Chapter 5 summarizes the full paper and presents future trends.

This book is largely based on the research work done in recent years by a team led by Dr. Zhifen Zhang and Dr. Guangrui Wen. These techniques provide valuable strategies for online monitoring of welding defects with non-destructive methods to improve welding productivity. Researchers, scientists and engineers working in the field of welding intelligence can benefit from this book.

Xi'an, China

Dr. Zhifen Zhang
Dr. Guangrui Wen
Dr. Wenjing Ren

Acknowledgements

Many thanks to Prof. Shanben Chen from Shanghai Jiao Tong University, who has been a great supervisor during and even after pursuing my Ph.D. degree. He has given me great support not only for my research work, but also for my personal career development and life. Wish him all well. Many thanks to Prof. Guangrui Wen from Xi'an Jiao Tong University, who has helped me a lot since I joined in his research team in XJTU from December 2015. I am looking forward to our better collaboration in future. Thanks for Dr. Wenjing Ren since the book cannot be finished without her former work in XJTU. Many thanks to several students from our research group, including Rui Qin, Jie Wang, Zhiwen Li, Zhengyao Du and Zijian Bai; they have been very supportive of my research and I have enjoyed working with them.

The work has been supported by National Natural Science Youth Foundation, No. 51605372; National Natural Science Foundation project No. 51775409, No. 52175433; Special support from China Postdoctoral Science Foundation, No. 2018T111052; China Postdoctoral Science Foundation, No. 2016M602805; National key research and development project, No. 2017YFF0210504, No. 2020YFB1710002; Equipment pre-research project, No. 6140004030116JW08001.

Contents

Chapter 1
Introduction

1.1 Background

Since Vasiliy Petrov discovered the phenomenon of continuous discharge in 1802 and proposed the potential application of electric arc in welding, arc has been widely used as a heat source in high-end fields such as aerospace, military equipment, and nuclear power facilities [1]. Especially in recent decades, with the continuous development of China's aerospace industry technology, welding technology has been widely used in the manufacturing of high-end aerospace equipment. According to statistics, welding technology accounts for 10% of the total aerospace workload and over 50% in the welding of components [2]. At present, welding technology has evolved from a single processing technique to a new interdisciplinary subject of modern science and technology, with the level of its development to a certain extent determining the overall level of advanced manufacturing in a country.

Traditional manual welding suffers from issues such as low efficiency, difficulty in ensuring quality, and unsafe production environment. This is because arc welding is a complex physical and chemical process involving multiple fields such as heat, electricity, magnetism, and force, which are coupled with each other. It has characteristics such as multivariable, randomness, and invisibility. The mechanism of occurrence and development of welding defects is ambiguous and complicated, with various types including internal defects, surface macroscopic defects, metallurgical defects, and industrial defects [3–6]. Welding defects greatly reduce the performance of welded structures. Light defects may require repair or rework, while severe defects may lead to direct scrap of the welded component. Undetected defects may deteriorate during equipment operation and lead to mechanical failures or even safety incidents. However, the application of automated and intelligent welding technology can effectively solve these problems and improve the reliability and stability of the welding process [7]. Smart manufacturing is the key development direction of China's manufacturing industry today, aiming to improve the efficiency, flexibility, and intelligence level of the manufacturing process [8]. Against this backdrop, automated and

Z. Zhang et al., *Multiple Sensing-Driven Feature Extraction and Defect Detection*,
https://doi.org/10.1007/978-981-96-4808-5_1

intelligent welding technology has become one of the important directions for the development of the manufacturing industry [9].

Automated welding technology achieves the automation of the welding process by introducing robots and automated equipment. This not only increases production efficiency but also reduces human errors, improves welding quality, and consistency. Moreover, automated systems can work in high-temperature, high-pressure, and hazardous environments, thus enhancing work safety and reducing the risk of injury to operators [10, 11]. With the continuous development of machine learning and artificial intelligence technologies, automated welding systems can gradually become intelligent. By learning and adjusting, they continuously optimize welding parameters, achieving higher levels of adaptability and intelligent control [12, 13].

Intelligent welding technology introduces more intelligent functions on the basis of automation. Through sensors, data acquisition, and analysis systems, it real-time monitors and analyzes key parameters and indicators during the welding process, as well as detects welding defects and quality issues [14–16]. Based on big data and artificial intelligence technology, intelligent welding systems can achieve fault prediction and prevention, adjust welding parameters in a timely manner to avoid defects, thereby improving welding quality and production efficiency. Additionally, intelligent systems can interact and integrate data with other manufacturing processes, realizing the informatization and intelligent management of the entire production process, further enhancing the overall level and competitiveness of the manufacturing industry [17].

The welding process is a complex physical and chemical process involving various factors such as electricity, light, heat, and force. Real-time monitoring of information such as arc sound, spectrum, current-voltage, visual images, weld pool oscillation, and weld pool temperature during the welding process can establish the connection between feature information and weld quality [18]. Real-time adjustment and optimization of welding parameters are the most effective technical means to fundamentally ensure welding quality [19]. In conclusion, welding digital technology provides necessary support for achieving automation and intelligence in production. The development and application of automated and intelligent welding technology will bring more efficient, safer, and smarter manufacturing processes to China's manufacturing industry, driving continuous progress in Chinese manufacturing.

1.2 Online Monitoring and Defects Detection for Arc Welding

The stability evaluation of welding process parameter detection and quantification is not only a prerequisite for monitoring welding equipment status and implementing automated feedback control in welding processes, but also an effective tool for predicting welding quality [20]. In-depth research on the stability evaluation technology of welding process parameter detection and quantification not only helps improve the automation level and quality management of welding processes but also provides important guidance for realizing intelligent manufacturing and promoting the digital transformation of welding processes [21]. Digital manufacturing technology is built on the basis of information technology, and achieves digital transformation of the production process through digital equipment, manufacturing environment, and corresponding digital resource support [22–24]. Given the demand for welding equipment status monitoring in the Internet of Things (IoT), studying the relationship between welding process signals and welding process stability and quality is of significant theoretical and practical importance [25, 26].

Compared to other physical information, the sound of the arc during the welding process can provide crucial information about welding quality and process status [15, 27]. It offers advantages such as immediacy (changes in arc sound can immediately reflect abnormalities during welding), intuitiveness (changes in arc sound can be directly perceived or analyzed through sound analysis software), and non-contact nature (monitoring arc sound does not require direct contact with the welding workpiece, avoiding interference with the welding process and maintaining its continuity and stability).

Additionally, spectral information during welding processes can provide detailed data on welding area temperature, material composition, and element distribution [28, 29]. Its advantages include: (1) Spectral analysis technology can provide high-precision temperature measurement and element analysis, assisting in accurately assessing the thermodynamic properties and chemical composition of the welding process. (2) Spectral information can be used to identify welding defects, monitor phase changes and microstructure evolution in metal structures, and evaluate the mechanical and corrosion resistance properties of welded joints. (3) Modern spectral analysis instruments can achieve high-speed data acquisition and real-time data processing, enabling online monitoring and control during the welding process, and timely adjustment of welding parameters. (4) With the help of wavelet analysis or Fourier transform, the temporal and spatial resolution of spectral information is high. (5) The spectrum contains rich information, including a large amount of process information about plasma morphology and physical characteristic parameters.

Therefore, the following sections will describe the mainstream welding quality online monitoring and defect detection technologies based on arc sound and arc spectral information, respectively.

1.2.1 Arc Sound-Based Sensing Technology

There exists a close relationship between arc sound and the welding process, with skilled welding engineers often able to identify transitions in droplet formation, the stability of the welding process, and the extent of spattering through arc sound. Variations in arc sound pressure and frequency are correlated with various factors, including the flickering of arc spots, oscillations of droplets and the weld pool, short circuits, arc re-ignition, and welding spatter [30, 31]. Additionally, arc sound exhibits varying degrees of correlation with multiple welding defects such as arc stability and the amount of spatter. Furthermore, arc sound signals offer advantages such as rapid response, rich information content, low susceptibility to interference, and easy installation of detectors [32]. With the rapid advancement of modern audio acquisition technology and signal processing noise reduction techniques, welding process monitoring technology has begun to utilize arc sound signals as an information source, attracting widespread attention from researchers.

A. **Arc sound characteristics and influencing factors**

The essence of sound signals is essentially a type of pressure wave generated by the vibration of an object in the air, which is modulated and propagated through the surrounding medium, and ultimately can be received by the human ear or a sound sensor [33–35]. Sound signals are mainly determined by the sound source (vibration source) and the propagation medium [36]. The sound collected during the arc welding process mainly comes from arc sound, shielding gas sound, welding equipment sound, and environmental noise. Compared with arc sound, the signal intensity of shielding gas and welding equipment sound is weak and can be negligible in practical applications of arc sound [37]. Therefore, in a broad sense, arc sound refers to the sound signal with the liquid metal caused by the arc impact on the weld pool and the vibration of the arc itself as the sound source during the welding process.

Currently, there is considerable research on the characteristics of arc sound generated during Gas Metal Arc Welding (GMAW) processes [38, 39]. Researchers at the University of Waterloo in Canada [40] have experimentally demonstrated that welding sound can effectively characterize the welding status, influencing the assessment of welding process quality. Cudina [41] and others have revealed the mechanisms of arc sound generation during GMAW short-circuit transition welding processes through theoretical analysis and experimental verification, and established the corresponding relationship between arc sound and current-voltage. The authors found that arc sound during GMAW short-circuit transition welding contains two different types of sound characteristics, namely "impulse noise" and "turbulent noise". Drouet [42] and others have analyzed the principles of arc sound generation during GMAW welding, and established the correspondence between arc sound and welding current and voltage, indicating that arc sound is closely related to changes in arc energy, metal materials, and arc vibration. Rostek [43] and others, through experimental verification, analyzed the direct impact of welding parameters such as welding voltage, wire feed speed, and shielding gas flow rate on arc sound during arc welding processes. At the same time, the authors also pointed out that the stability and distribution of shielding gas, contamination of welding materials, and deviations in the morphology of welding joints also affect the generation of arc sound.

Saini [37] conducted time-frequency analysis on collected sound signals. In the time domain, they extracted multiple parameters such as amplitude, zero crossing rate, and waveform factor, and analyzed the relationship between these feature parameters and the transition state of molten droplets. Reference [44], focusing on the GMAW welding process, measured the sound pressure of different materials and simultaneously detected the time-domain and frequency-domain characteristics of arc sound, finding that arc sound exhibits significant peaks at higher frequencies (above 5 kHz). Reference [45] studied arc sound during short-circuit transition in the GMAW process, demonstrating that the changes in arc sound during short-circuit transition are mainly due to the occurrence of arc extinction and re-ignition phenomena. Horvat [46] and others focused on Pulsed Gas Melt Arc Welding (P-GMAW), pointing out that arc sound consists of splatter sound, molten droplet transition sound, and re-excited sound in GMAW.

Lv [30] have conducted systematic research on arc sound during welding processes and its application in welding defect detection. They point out that arc sound generation consists of two parts: firstly, the variation of arc energy in the near-conical transparent cavity formed by the shielding gas, electrode, and material excites the sound source, which is then modulated to form arc sound. Ebrahimi et al. [47] established a resonant model of arc sound and weld pool, analyzing the influence of weld pool oscillation on arc sound signals. The analysis showed that the morphology of the weld pool plays a decisive role in determining the characteristic frequencies of arc sound. Zhang et al. [31] analyzed the response of arc sound to welding conditions in the time domain, frequency domain, and time-frequency domain. By comparing the synchronous acquisition of current, voltage, and arc sound signals during GTAW welding processes, they revealed a high correlation between arc sound and arc energy variation.

B. **Processing of arc sound and identification of welding penetration status**

The objective of arc sound signal research is to analyze the variations in signals corresponding to different melt-through states and defect characteristics, and to utilize this information to establish melt-through state prediction models and control strategies for effective control of welding quality [48]. In plasma arc piercing welding processes, melt pool oscillation is the main source of low-frequency components in sound signals. Pashazadeh et al. [49] obtained sound signals during laser welding processes and conducted analyses of sound dynamics in the time and frequency domains through synchronous wavelet compression analysis. Lv et al. [50] solved the frequency information of arc sound through wavelet analysis and wavelet packet analysis, and reduced high-frequency noise interference through noise reduction techniques. Ren et al. [51] proposed a monitoring architecture that combines arc sound time-frequency maps and improved convolutional neural networks, and applied it to different penetration depths in GTAW.

Statistical features of arc sound signals in the time and frequency domains, such as root mean square, mean amplitude, kurtosis, and energy values in different frequency bands, are sensitive to welding state signals [52]. Commonly used speech recognition features such as Linear Predictive Coding Coefficients (LPCC) [53], Power Spectral Density (PSD) [54], Mel-frequency Cepstral Coefficients (MFCC) [55], and others [56] have also been used to process arc sound signals. In addition, methods such as Empirical Mode Decomposition (EMD), Principal Component Analysis (PCA), and statistical analysis are commonly used for feature extraction and dimensionality reduction in arc sound signals [50, 57]. The extracted arc sound features can be used for monitoring welding state stability, and when combined with artificial intelligence recognition models, can achieve weld penetration state detection and control. Especially in recent years, artificial intelligence models have been used in several fields [58–60]. Lv [27, 50, 61], through wavelet decomposition of arc sound signals from GTAW aluminum alloy welding, extracted time-domain and frequency-domain features of signals at different scales after decomposition, and achieved weld penetration state recognition and control by establishing an Artificial Neural Network (ANN). Zhang et al. [31] compared the Fisher distance and PCA methods for arc sound feature extraction, and used the arc sound frequency features obtained by different methods to train a Support Vector Machine (SVM) recognition model, ultimately achieving recognition of three weld penetration states: unwelded, normal, and burn-through. Prior to Support Vector Machine classification analysis, dynamics of sound in the time and frequency domains were analyzed. Their conclusion was that features extracted through synchronous wavelet compression analysis can be used to classify sound signals of different penetration states.

In summary, existing research has conducted a significant amount of work on the generation mechanism, preprocessing, and feature extraction of arc sound signals. It has demonstrated the close relationship between arc sound signals and weld penetration state, and combined with classical artificial intelligence recognition models, achieved intelligent detection of weld penetration state based on arc sound signal features [31, 48]. However, due to the complexity and difficulty in separating the sources of acoustic signals during welding processes, various factors such as the transition of molten metal droplets, release of internal tension, shielding gas flow, and molten pool flow can generate sound. Moreover, acoustic signals are easily affected by external environmental factors and other interferences, impacting measurement accuracy. These factors make it challenging to establish a definitive relationship between sound signals and welding stability and quality, thereby significantly limiting the practical application of sound signals in production. In addition, with the development of artificial intelligence technology, combining sound signals with intelligent models is also a potential direction for the future [13, 62].

1.2.2 Arc Spectrum-Based Sensing Technology

Metal or shielding gas is excited and ionized under the action of thermal effects and free electrons in the arc, forming a gaseous plasma composed of positive and negative ions. Plasma spectroscopy refers to the light waves emitted during the transition of particles from high energy levels to low energy levels while in an excited state, and it can be used to calculate parameters such as plasma composition, electron density, and plasma temperature [63, 64]. These parameters are closely related to welding quality. Furthermore, spectroscopy has advantages such as high information content and strong sensitivity. Compared to other signals such as images, arc spectroscopy has certain advantages in detecting internal defects such as burning and porosity in welding processes [65, 66].

A. **Application status of arc spectrum in welding quality inspection**

In recent years, with the development of electronics and mechanics towards miniaturization, advanced sensing technology has made significant progress. Small spectrometers with light weight and fast detection speed are continuously updated and upgraded. Many scholars have applied them to the field of welding, collecting arc welding or laser welding process arc spectrum information to monitor and improve welding quality, and have made certain progress [67–69]. At present, based on the type of spectrum information adopted, research in this area mainly focuses on two directions: based on arc plasma temperature (Te) and based on arc spectrum intensity.

Li [70], as an early researcher in arc spectroscopy, pointed out that arc spectrum information can reflect the state and changes of the arc. He summarized and analyzed 12 equations in the theories of plasma physics and spectroscopy, which can be used to analyze the characteristics of spectrum signals, providing a theoretical basis for arc spectrum research. Mirapeix [71] calculated the arc temperature (Te) in the TIG welding spectrum of stainless steel plates, using it for online detection of welding defects. Experimental results showed that when the argon flow rate changed, leading to defects in the weld bead, there was a significant abrupt change in arc temperature (Te). However, welding defects have various sources, and not all defects in welding can be detected by arc temperature (Te) monitoring alone; it requires combined analysis with other signals. Cobo et al. [72] used the arc electron temperature (Te) curve to detect welding quality in tube plate welding of ENSA steam generators. Experiments showed that the Te curve at the weld crater caused by the collapse of the molten pool showed an overall decrease, the Te curve fluctuated significantly at the tungsten clamp of the weld bead, and the Te curve fluctuated dramatically at unfused defects. Although the Te curve exhibits different characteristics when welding defects occur, it is somewhat difficult to solely distinguish between different weld defects using the Te curve [73].

Research on arc plasma temperature currently focuses on studying the three-dimensional temperature field of the arc [74, 75]. By collecting spectra from different positions of the arc using a spectrometer, the arc plasma temperature can be calculated. Based on the Abel transform, the arc temperature field at different positions can be obtained, thereby diagnosing and studying the physical characteristics of the welding arc. This is helpful for improving welding processes, especially for composite welding. For example, Chen [76] used the Boltzmann method to study the distribution of arc electron temperature in multi-wire GMAW welding processes.

However, the calculation method of plasma temperature requires plasma to satisfy local thermodynamic equilibrium and optical thin conditions [70]. Its calculation formula itself belongs to ideal formulas and requires high experimental precision and spectroscopic measurement accuracy. Moreover, due to the inherent fluctuations, randomness, complexity, and uncertain interference in the welding process, it often cannot accurately reflect the changes in the welding process [77]. Arc spectrum intensity directly reflects the intensity of arc radiation and can more directly and comprehensively reflect the dynamic changes in the welding process without introducing secondary errors. Therefore, it is more suitable for practical welding applications [78].

Huang [79] studied the relationship between the intensity of spectral lines and porosity, and applied different amounts of pollutants on the weld to create porosity defects of different sizes. Spectral signal features were extracted through a series of signal processing methods, and the distribution pattern of porosity during the welding process was quantitatively analyzed. Zhang [80] analyzed the spectra during the arc additive process. Mirapeix [81] studied the aluminum atomic spectrum lines in laser welding of boron steel protected by aluminum-silicon coatings, and analyzed and compared the tensile properties and fracture positions of the welds. It was found that the intensity of aluminum spectrum lines in the spectral signal can reflect the tensile strength of the welds, and it was also observed that specimens with high aluminum spectrum line intensity during the welding process were prone to crack defects. Song [82] conducted research on aluminum atomic spectrum lines in laser additive manufacturing processes, extracted the intensity ratio and integrated intensity of aluminum spectrum lines, and proposed a Support Vector Regression (SVR) method with adjustable parameters to predict component concentrations under different parameters, providing a method for real-time component monitoring in laser additive manufacturing processes.

From the above analysis, it can be seen that the arc plasma temperature feature requires high experimental precision and may introduce secondary errors. Even if the above requirements are met, a single arc plasma temperature feature alone is difficult to detect multiple welding defects. In the study of welding spectrum line intensity, most research focuses on extracting information from a single spectral line or a specific spectral frequency band, analyzing a single welding defect, which not only overlooks the richness of arc spectra but also fails to meet the requirements of welding inspection in industrial settings. Further research is needed to fully explore the spectral lines in the spectrum that are relevant to multiple typical welding defects.

B. **Feature selection and dimension reduction of arc spectrum**

Arc spectra contain information about multiple atomic spectral lines within the arc, and a single atomic spectral line may have multiple lines. While this richness of information is valuable, it also includes a lot of redundant and irrelevant information, which increases the difficulty of data processing and can affect the accuracy and efficiency of modeling. The key challenge in current arc spectrum research is how to establish a welding defect recognition model with fewer effective feature information. Through feature selection or dimensionality reduction, retaining important information from the original spectral data while using a small number of features for training and prediction in machine learning models can improve the model's recognition accuracy and speed, addressing the critical issue of welding online detection [83, 84].

There are mainly two methods for reducing feature dimensions and addressing the curse of dimensionality [65]. One is feature selection, which involves finding a subset containing important features from the original feature set through various evaluation methods. This process does not generate new features. The mainstream approaches for feature selection include filter methods, wrapper methods, and embedded methods. The other method is feature dimensionality reduction, which involves using a mapping method to recombine or map the original features into a different dimensional space. The main methods for feature dimensionality reduction include principal component analysis (PCA) [84], independent component analysis (ICA) [29], and locally linear embedding (LLE) [79].

Bruncko [85] conducted research on the relationship between laser welding plasma and welding parameters, analyzing the impact of welding processes on plasma. With the increase in welding power, the intensity of infrared and blue-purple wavelength signals in the spectral data linearly increases. Li [86] studied the changes in arc spectra during different states of gas tungsten arc welding (GTAW). Based on this research, five special spectral regions with high signal-to-noise ratios were artificially extracted to detect various welding defects caused by different interfering factors. Garcia-Allende et al. [87] extracted 50 spectral lines from arc spectral signals using the SFFS algorithm. They then classified four scenarios—normal welding, reduced shielding gas, notching, and low welding current—using a neural network, achieving good classification results. Real et al. [63] employed PCA and ICA for correlation analysis and redundancy removal of spectral data. They ultimately divided the spectral information into three independent components, each corresponding to different types of plasma generation mechanisms.

In the field of welding manufacturing, research on arc spectral information has been initiated and has achieved certain results. There has also been some development in dimensionality reduction of spectral signals. However, in practical applications, there are still some issues with the stability, real-time performance, and interpretability of spectral feature reduction methods. Currently, in terms of feature selection, manually selecting the interested spectral regions and the susceptibility of the Sequential Forward Floating Selection (SFFS) method to human bias are limiting factors. Although PCA and ICA have achieved preliminary selection of welding spectral features, they have not yet extracted the optimal feature subset. In addition, these are data-driven dimensionality reduction methods that sacrifice the interpretability of the data, making it difficult to analyze the correspondence between spectral features and weld defects.

C. **Defect recognition model for arc spectrum**

By establishing the relationship between spectral signals and defects, spectral signals can be used to detect welding defects. With the rise of the machine learning trend, various machine learning algorithms have shown outstanding performance in classification and regression problems [88, 89]. Scholars have continuously attempted to establish welding defect recognition models based on spectral signals. Among them, neural networks are mathematical models that simulate the structure of the human brain and are highly efficient computational models used for approximating or estimating target functions [90]. For instance, Kim et al. [91] developed a back propagation (BP) neural network model to predict weld bead width in steel MAG welding processes using four welding parameters as input layer inputs. Huang [92] used a genetic algorithm-based support vector machine model to quantitatively identify porosity defects in aluminum alloy welding processes by employing the ratio of argon spectrum to hydrogen spectrum features. Mirapeix et al. [93] established a BP neural network with two hidden layers using multiple element spectral line information to identify four different welding states, including lack of penetration and low current, in stainless steel thin plate welding. In addition, some scholars have applied fuzzy theory to spectral feature-based welding defect detection [94]. Fuzzy theory is mainly used for models with complex mathematical characteristics that are not quantified and possess non-precise, non-linear information processing capabilities. Naso et al. [95] collected ultraviolet, infrared, and visible spectra during welding processes, along with computed arc temperature, to classify weld penetration status and six types of welding defects.

Random forests, derived from perfecting and improving decision trees, have been increasingly applied in various fields due to their excellent classification performance and inherent feature importance indicators [28]. In the field of pattern recognition, Bonissone [96] proposed a fuzzy random forest algorithm based on the random forest model and fuzzy logic theory, achieving high classification accuracy. Chang [97] utilized a random forest classifier to segment depth images of faces, demonstrating the effectiveness of the proposed method. Wu [98] employed random forests to establish a rating evaluator, realizing the grading evaluation of tool wear. Additionally, random forests have been widely applied in hyperspectral imagery [99, 100] and medical fields [101, 102]. In terms of feature selection, Yao et al. [103] used classification progress as an evaluation metric and employed the sequential backward selection method for feature selection. Zhang [104] addressed high-dimensional imbalanced data by utilizing the weighted average of decision tree weights and feature importance indicators to determine the final feature importance sequence.

Currently established mathematical models for defect recognition require comprehensive and abundant training data. In welding industrial environments, obtaining various defect data samples is challenging [105]. Furthermore, there are many nonlinear factors in the welding process [4]. To achieve better recognition results, the models currently established are generally aimed at single defects. Even for multi-defect recognition models, there are drawbacks such as complex algorithms, slow computational speed, and low recognition efficiency, making it difficult to meet the requirements of online detection [106, 107]. In the future, developing interpretable artificial intelligence technology is also an important direction [108–110].

References

1. S. Chen and N. Lv, "Research evolution on intelligentized technologies for arc welding process," *Journal of manufacturing processes,* vol. 16, no. 1, pp. 109–122, 2014.
2. T. Mohandas, "Welding Technologies in Aerospace Applications," *Aerospace Materials and Material Technologies: Volume 2: Aerospace Material Technologies,* pp. 65–83, 2017.
3. B. R. Somers and A. W. Pense, "Welding failure analysis," *Materials characterization,* vol. 33, no. 3, pp. 295–309, 1994.

4. A. S. Madhvacharyula, A. V. S. Pavan, S. Gorthi, S. Chitral, N. Venkaiah, and D. V. Kiran, "In situ detection of welding defects: A review," *Welding in the World,* vol. 66, no. 4, pp. 611–628, 2022.
5. Z. ZHANG, Z. YANG, W. REN, and G. WEN, "Condition detection in Al alloy welding process based on deep mining of arc spectrum," *Transactions of the China Welding Institution,* vol. 40, no. 1, pp. 19–25, 2019.
6. J. Wang, Z. Zhang, R. Qin, and G. Wen, "Online identification of burn-through and weld deviation in sheet lap MIG welding based on YOLOv5," *Measurement Science and Technology,* vol. 35, no. 2, p. 025119, 2023.
7. S. Chen *et al.*, "Intelligent technologies for robotic welding," in *Robotic welding, intelligence and automation*, 2004: Springer, pp. 123–143.
8. R. Qin *et al.*, "On-line evaluation and monitoring technology for material surface integrity in laser shock peening–A review," *Journal of Materials Processing Technology,* vol. 313, p. 117851, 2023.
9. B. Wang, S. J. Hu, L. Sun, and T. Freiheit, "Intelligent welding system technologies: State-of-the-art review and perspectives," *Journal of Manufacturing Systems,* vol. 56, pp. 373–391, 2020.
10. Y. Huang and S. Chen, *Key Technologies of Intelligentized Welding Manufacturing.* Springer, 2020.
11. Z. Zhang, Y. Huang, R. Qin, W. Ren, and G. Wen, "XGBoost-based on-line prediction of seam tensile strength for Al-Li alloy in laser welding: Experiment study and modelling," *Journal of Manufacturing Processes,* vol. 64, pp. 30–44, 2021.
12. Z. Zhang, G. Wen, and S. Chen, "On-line monitoring and defects detection of robotic arc welding: A review and future challenges," *Transactions on Intelligent Welding Manufacturing: Volume II No. 4 2018,* pp. 3–28, 2019.
13. Z. Zhang *et al.*, "Deep learning-based monitoring of surface residual stress and efficient sensing of AE for laser shock peening," *Journal of Materials Processing Technology,* vol. 303, p. 117515, 2022.
14. Q. Guo *et al.*, "Progress, challenges and trends on vision sensing technologies in automatic/intelligent robotic welding: State-of-the-art review," *Robotics and Computer-Integrated Manufacturing,* vol. 89, p. 102767, 2024.
15. Z. Zhang and S. Chen, "Real-time seam penetration identification in arc welding based on fusion of sound, voltage and spectrum signals," *Journal of Intelligent Manufacturing,* vol. 28, pp. 207–218, 2017.
16. J. Huang *et al.*, "Lightweight neural network architecture for pipeline weld crack leakage monitoring using acoustic emission," *IEEE Transactions on Instrumentation and Measurement,* 2023.
17. S. Chen, "On the key technologies of intelligentized welding robot," in *Robotic welding, intelligence and automation*: Springer, 2007, pp. 105–115.
18. F. Xu, Y. Xu, H. Zhang, and S. Chen, "Application of sensing technology in intelligent robotic arc welding: A review," *Journal of Manufacturing Processes,* vol. 79, pp. 854–880, 2022.
19. Z. Zhang, H. Chen, Y. Xu, J. Zhong, N. Lv, and S. Chen, "Multisensor-based real-time quality monitoring by means of feature extraction, selection and modeling for Al alloy in arc welding," *Mechanical systems and signal processing,* vol. 60, pp. 151–165, 2015.
20. Z. Zhang, X. Chen, H. Chen, J. Zhong, and S. Chen, "Online welding quality monitoring based on feature extraction of arc voltage signal," *The International Journal of Advanced Manufacturing Technology,* vol. 70, pp. 1661–1671, 2014.
21. I.-S. Kim, Y.-J. Jeong, C.-W. Lee, and P. K. Yarlagadda, "Prediction of welding parameters for pipeline welding using an intelligent system," *The International Journal of Advanced Manufacturing Technology,* vol. 22, pp. 713–719, 2003.
22. J. Huang and R. Qin, "Elman neural network considering dynamic time delay estimation for short-term forecasting of offshore wind power," *Applied Energy,* vol. 358, p. 122671, 2024.
23. Z. Zhang *et al.*, "Rebalancing Mel-frequency Cepstrum and parallel fusion model for surface hardness monitoring of laser shock peening using acoustic emission," *Mechanical Systems and Signal Processing,* vol. 223, p. 111912, 2025.

24. R. Qin *et al.*, "Accelerable adaptive cepstrum and L2-Dual Net for acoustic emission-based quality monitoring in laser shock peening," *Journal of Manufacturing Systems,* vol. 77, pp. 301–319, 2024.
25. Q. Liu, C. Chen, and S. Chen, "Key technology of intelligentized welding manufacturing and systems based on the internet of things and multi-agent," *Journal of Manufacturing and Materials Processing,* vol. 6, no. 6, p. 135, 2022.
26. J. Wang *et al.*, "On-line defect recognition of MIG lap welding for stainless steel sheet based on weld image and CMT voltage: Feature fusion and attention weights visualization," *Journal of Manufacturing Processes,* vol. 108, pp. 430–444, 2023.
27. N. Lv, Y. Xu, S. Li, X. Yu, and S. Chen, "Automated control of welding penetration based on audio sensing technology," *Journal of Materials Processing Technology,* vol. 250, pp. 81–98, 2017.
28. Z. Zhang, Z. Yang, W. Ren, and G. Wen, "Random forest-based real-time defect detection of Al alloy in robotic arc welding using optical spectrum," *Journal of Manufacturing Processes,* vol. 42, pp. 51–59, 2019.
29. Z. Zhang, Y. Huang, R. Qin, Z. Lei, and G. Wen, "Real-time measurement of seam strength using optical spectroscopy for Al–Li alloy in laser beam welding," *IEEE Transactions on Instrumentation and Measurement,* vol. 70, pp. 1–10, 2021.
30. N. Lv and S. Chen, *Key technologies of intelligentized welding manufacturing: welding arc acoustic sensing and monitoring technology*. Springer, 2020.
31. Z. Zhang, G. Wen, and S. Chen, "Audible sound-based intelligent evaluation for aluminum alloy in robotic pulsed GTAW: mechanism, feature selection, and defect detection," *IEEE Transactions on Industrial Informatics,* vol. 14, no. 7, pp. 2973–2983, 2017.
32. Z. Zhang, R. Qin, Y. Yuan, W. Ren, Z. Yang, and G. Wen, "Acoustic emission-based weld crack in-situ detection and location using WT-TDOA," in *Transactions on Intelligent Welding Manufacturing: Volume III No. 4 2019*, 2021: Springer, pp. 49–73.
33. J. Huang *et al.*, "Acoustic emission technology-based multifractal and unsupervised clustering on crack damage monitoring for low-carbon steel," *Measurement,* vol. 217, p. 113042, 2023.
34. Y. Yu *et al.*, "Acoustic emission-based weld crack leakage monitoring via FGI and MCCF-CondenseNet convolutional neural network," *NDT & E International,* vol. 148, p. 103232, 2024.
35. S. Zhang *et al.*, "Intra-layer and inter-layer monitoring of laser powder bed fusion defects based on airborne acoustic and gn-Res model: pore and deformation," *Virtual and Physical Prototyping,* vol. 19, no. 1, p. e2360699, 2024.
36. R. Qin *et al.*, "An adaptive cepstrum feature representation method with variable frame length and variable filter banks for acoustic emission signals," *Mechanical Systems and Signal Processing,* vol. 208, p. 111031, 2024.
37. D. Saini and S. Floyd, "An investigation of gas metal arc welding sound signature for on-line quality control," *WELDING JOURNAL-NEW YORK-,* vol. 77, pp. 172-s, 1998.
38. L. Nickels, "AM and aerospace: an ideal combination," *Metal Powder Report,* vol. 70, no. 6, pp. 300–303, 2015.
39. S. M. Thompson, L. Bian, N. Shamsaei, and A. Yadollahi, "An overview of Direct Laser Deposition for additive manufacturing; Part I: Transport phenomena, modeling and diagnostics," *Additive Manufacturing,* vol. 8, pp. 36–62, 2015.
40. J. Tam, "Methods of characterizing gas-metal arc welding acoustics for process automation," University of Waterloo, 2005.
41. M. Čudina and J. Prezelj, "Evaluation of the sound signal based on the welding current in the gas—metal arc welding process," *Proceedings of the Institution of Mechanical Engineers, Part C: Journal of Mechanical Engineering Science,* vol. 217, no. 5, pp. 483–494, 2003.
42. M. G. Drouet and F. Nadeau, "Acoustic measurement of the arc voltage applicable to arc welding and arc furnaces," *Journal of Physics E: Scientific Instruments,* vol. 15, no. 3, p. 268, 1982.
43. W. Rostek, "INVESTIGATION ON THE CONNECTION BETWEEN THE WELDING PROCESS AND AIRBORNE NOISE EMISSION IN GAS-SHIELDED METAL-ARC WELDING," 1990.

44. E. Cayo and S. A. Alfaro, "Weld transference modes identification through sound pressure level in GMAW process," *Journal of Achievements in Materials and Manufacturing Engineering,* vol. 29, no. 1, pp. 57–62, 2008.
45. Y. Cui, Y. Shi, T. Zhu, and S. Cui, "Welding penetration recognition based on arc sound and electrical signals in K-TIG welding," *Measurement,* vol. 163, p. 107966, 2020.
46. J. Horvat, J. Prezelj, I. Polajnar, and M. Čudina, "Monitoring gas metal arc welding process by using audible sound signal," *Strojniški vestnik-Journal of Mechanical Engineering,* vol. 57, no. 3, pp. 267–278, 2011.
47. A. Ebrahimi, C. R. Kleijn, M. J. Hermans, and I. M. Richardson, "The effects of process parameters on melt-pool oscillatory behaviour in gas tungsten arc welding," *Journal of Physics D: Applied Physics,* vol. 54, no. 27, p. 275303, 2021.
48. J. Huang *et al.*, "Residual swin transformer-based weld crack leakage monitoring of pressure pipeline," *Welding in the World,* vol. 68, no. 4, pp. 879–891, 2024.
49. H. Pashazadeh, Y. Gheisari, and M. Hamedi, "Statistical modeling and optimization of resistance spot welding process parameters using neural networks and multi-objective genetic algorithm," *Journal of Intelligent Manufacturing,* vol. 27, pp. 549–559, 2016.
50. N. Lv, Y. Xu, Z. Zhang, J. Wang, B. Chen, and S. Chen, "Audio sensing and modeling of arc dynamic characteristic during pulsed Al alloy GTAW process," *Sensor Review,* vol. 33, no. 2, pp. 141–156, 2013.
51. W. Ren, G. Wen, B. Xu, and Z. Zhang, "A novel convolutional neural network based on time–frequency spectrogram of arc sound and its application on GTAW penetration classification," *IEEE Transactions on Industrial Informatics,* vol. 17, no. 2, pp. 809–819, 2020.
52. C. Chen, R. Xiao, H. Chen, N. Lv, and S. Chen, "Arc sound model for pulsed GTAW and recognition of different penetration states," *The International Journal of Advanced Manufacturing Technology,* vol. 108, pp. 3175–3191, 2020.
53. J. Wang, Y. Zuo, Y. Huang, B. Yang, and S. Pan, "Arc sound recogniting penetration state using LPCC features," in *Robotic Welding, Intelligence and Automation: RWIA'2010*: Springer, 2011, pp. 229–233.
54. Y. Wang and Q. Chen, "On-line quality monitoring in plasma-arc welding," *Journal of Materials Processing Technology,* vol. 120, no. 1–3, pp. 270–274, 2002.
55. Z. Zhang, R. Qin, G. Li, Z. Liu, G. Wen, and W. He, "Online evaluation of surface hardness for aluminum alloy in LSP using modal acoustic emission," *IEEE Transactions on Instrumentation and Measurement,* vol. 71, pp. 1–10, 2022.
56. Z. Zhang, R. Qin, G. Li, Z. Du, G. Wen, and W. He, "A novel approach for surface integrity monitoring in high-energy nanosecond-pulse laser shock peening: acoustic emission and hybrid-attention CNN," *IEEE Transactions on Industrial Informatics,* vol. 19, no. 3, pp. 2802–2813, 2022.
57. M. Yusof, M. Kamaruzaman, M. Ishak, and M. Ghazali, "Porosity detection by analyzing arc sound signal acquired during the welding process of gas pipeline steel," *The International Journal of Advanced Manufacturing Technology,* vol. 89, pp. 3661–3670, 2017.
58. J. Huang *et al.*, "An incremental learning framework for pipeline weld crack damage identification and leakage rate prediction," *Engineering Applications of Computational Fluid Mechanics,* vol. 18, no. 1, p. 2406256, 2024.
59. Z. Zhang, Z. Du, R. Qin, G. Li, and G. Wen, "Surface hardness monitoring of laser shock Peening: Acoustic emission and key frame selection," *Measurement,* vol. 199, p. 111560, 2022.
60. J. Huang *et al.*, "Dual channel visible graph convolutional neural network for microleakage monitoring of pipeline weld homalographic cracks," *Computers in Industry,* vol. 164, p. 104193, 2025.
61. N. Lv, J. Zhong, H. Chen, T. Lin, and S. Chen, "Real-time control of welding penetration during robotic GTAW dynamical process by audio sensing of arc length," *The International Journal of Advanced Manufacturing Technology,* vol. 74, pp. 235–249, 2014.
62. J. Huang *et al.*, "Interpretable real-time monitoring of pipeline weld crack leakage based on wavelet multi-kernel network," *Journal of Manufacturing Systems,* vol. 72, pp. 93–103, 2024.

63. E. Real, J. Mirapeix, O. Conde, R. Ruiz-Lombera, A. Cobo, and J. Lopez-Higuera, "Species discrimination in plasma welding spectra by means of principal and independent component analysis," in *23rd International Conference on Optical Fibre Sensors*, 2014, vol. 9157: SPIE, pp. 49–52.
64. Z. Zhang, H. Yu, N. Lv, and S. Chen, "Real-time defect detection in pulsed GTAW of Al alloys through on-line spectroscopy," *Journal of Materials Processing Technology,* vol. 213, no. 7, pp. 1146–1156, 2013.
65. Z. Zhang, L. Zhang, and G. Wen, "Study of inner porosity detection for Al-Mg alloy in arc welding through on-line optical spectroscopy: Correlation and feature reduction," *Journal of Manufacturing Processes,* vol. 39, pp. 79–92, 2019.
66. Z. Li *et al.*, "A novel approach of online monitoring for laser powder bed fusion defects: Air-borne Acoustic Emission and Deep Transfer Learning," *Journal of Manufacturing Processes,* vol. 102, pp. 579–592, 2023.
67. M. S. Węglowski, "Measurement of arc light spectrum in the MAG welding method," *Metrol. Meas. Syst,* vol. 16, no. 1, pp. 143–159, 2009.
68. M. S. Węglowski, "Monitoring of arc welding process based on arc light emission," *Welding Processes,* pp. 8s-13s, 2012.
69. M. Weglowski and M. Kepinska, "An experimental study on the relationship between welding parameters and the arc light spectrum in TIG welding," *WELDING IN THE WORLD-LONDON-,* vol. 51, no. I, p. 291, 2007.
70. J.-y. LI, Y.-l. SONG, L.-j. YANG, and H.-t. XUE, "Basic Theory and Method of Welding Arc Spectrum Informatin," *Transactions of the China Welding Institution,* no. 6, pp. 5–8, 2002.
71. J. Mirapeix, A. Cobo, O. Conde, C. Jaúregui, and J. López-Higuera, "Real-time arc welding defect detection technique by means of plasma spectrum optical analysis," *Ndt & E International,* vol. 39, no. 5, pp. 356–360, 2006.
72. A. Cobo, J. Mirapeix, F. Linares, J. A. Piney, D. Solana, and J. M. Lopez-Higuera, "Spectroscopic sensor system for quality assurance of the tube-to-tubesheet welding process in nuclear steam generators," *IEEE Sensors Journal,* vol. 7, no. 9, pp. 1219–1224, 2007.
73. U. Reisgen, C. Otten, and J. Schönberger, "Investigations about the influence of the time–temperature curve on the formation of intermetallic phases during electron beam welding of steel–aluminium material combinations," *Welding in the World,* vol. 58, pp. 443–454, 2014.
74. P. Freton, J. Gonzalez, and A. Gleizes, "Comparison between a two-and a three-dimensional arc plasma configuration," *Journal of Physics D: Applied Physics,* vol. 33, no. 19, p. 2442, 2000.
75. K. F. He, X. J. Li, J. G. Wu, and Q. Li, "Three-dimensional temperature field numerical simulation of twin-arc high-speed submerged arc welding process based on ANSYS," *Advanced Materials Research,* vol. 216, pp. 188–193, 2011.
76. X. Chen, H. Xue-ming, Y. Ding-jian, M. Xiao-li, L. Fang, and H. Ye, "Study of the Effect of Interference during Multi-Wire GMAW Based on Spectral Diagnosis Technique," *Spectroscopy and Spectral Analysis,* vol. 38, no. 7, pp. 1993–1997, 2018.
77. Y. N. Seregin and S. Kurashkin, "Modeling the thermal process using the temperature functional by electron beam welding," in *IOP Conference Series: Materials Science and Engineering*, 2020, vol. 734, no. 1: IOP Publishing, p. 012003.
78. T. Sibillano *et al.*, "Study on the correlation between plasma electron temperature and penetration depth in laser welding processes," *Physics Procedia,* vol. 5, pp. 429–436, 2010.
79. Y. Huang, Y. Yuan, L. Yang, Z. Zhang, and S. Hou, "A study on porosity in gas tungsten arc welded aluminum alloys using spectral analysis," *Journal of Manufacturing Processes,* vol. 57, pp. 334–343, 2020.
80. C. Zhang, M. Gao, C. Chen, and X. Zeng, "Spectral diagnosis of wire arc additive manufacturing of Al alloys," *Additive Manufacturing,* vol. 30, p. 100869, 2019.
81. J. Mirapeix, E. Vila, J. Valdiande, A. Riquelme, M. Garcia, and A. Cobo, "Real-time detection of the aluminium contribution during laser welding of Usibor1500 tailor-welded blanks," *Journal of Materials Processing Technology,* vol. 235, pp. 106–113, 2016.

82. L. Song, W. Huang, X. Han, and J. Mazumder, "Real-time composition monitoring using support vector regression of laser-induced plasma for laser additive manufacturing," *IEEE Transactions on Industrial Electronics,* vol. 64, no. 1, pp. 633–642, 2016.
83. Z. Zhang and S. Chen, "Data-driven feature selection for multisensory quality monitoring in arc welding," in *International Conference on Robotic Welding, Intelligence and Automation,* 2014: Springer, pp. 401–410.
84. Z. Zhang, W. Ren, Z. Yang, and G. Wen, "Real-time seam defect identification for Al alloys in robotic arc welding using optical spectroscopy and integrating learning," *Measurement,* vol. 156, p. 107546, 2020.
85. J. Bruncko, F. Uherek, D. Chorvat Jr, M. Michalka, and P. Fodrek, "Spectroscopic investigation of laser-induced plasma in laser beam welding," in *Seventh International Conference on Laser and Laser-Information Technologies*, 2002, vol. 4644: SPIE, pp. 109–115.
86. L. Zhiyong, W. Bao, and D. Jingbin, "Detection of GTA welding quality and disturbance factors with spectral signal of arc light," *Journal of materials processing technology,* vol. 209, no. 10, pp. 4867–4873, 2009.
87. P. Garcia-Allende, J. Mirapeix, O. Conde, A. Cobo, and J. Lopez-Higuera, "Spectral processing technique based on feature selection and artificial neural networks for arc-welding quality monitoring," *Ndt & E International,* vol. 42, no. 1, pp. 56–63, 2009.
88. R. Qin *et al.*, "Acoustic emission for surface quality monitoring in laser shock peening via dual-feature fusion convolution neural network," *Optics & Laser Technology,* vol. 164, p. 109505, 2023.
89. R. Qin *et al.*, "Surface stress monitoring of laser shock peening using AE time-scale texture image and multi-scale blueprint separable convolutional networks with attention mechanism," *Expert Systems with Applications,* vol. 224, p. 120018, 2023.
90. S. Abe, *Neural networks and fuzzy systems: theory and applications.* Springer Science & Business Media, 2012.
91. I.-S. Kim, J.-S. Son, S.-H. Lee, and P. K. Yarlagadda, "Optimal design of neural networks for control in robotic arc welding," *Robotics and computer-integrated manufacturing,* vol. 20, no. 1, pp. 57–63, 2004.
92. Y. Huang, D. Wu, Z. Zhang, H. Chen, and S. Chen, "EMD-based pulsed TIG welding process porosity defect detection and defect diagnosis using GA-SVM," *Journal of Materials Processing Technology,* vol. 239, pp. 92–102, 2017.
93. J. Mirapeix, P. García-Allende, A. Cobo, O. Conde, and J. López-Higuera, "Real-time arc-welding defect detection and classification with principal component analysis and artificial neural networks," *NDT & e International,* vol. 40, no. 4, pp. 315–323, 2007.
94. H.-J. Zimmermann, *Fuzzy set theory—and its applications.* Springer Science & Business Media, 2011.
95. D. Naso, B. Turchiano, and P. Pantaleo, "A fuzzy-logic based optical sensor for online weld defect-detection," *IEEE transactions on Industrial Informatics,* vol. 1, no. 4, pp. 259–273, 2005.
96. P. Bonissone, J. M. Cadenas, M. C. Garrido, and R. A. Díaz-Valladares, "A fuzzy random forest," *International Journal of Approximate Reasoning,* vol. 51, no. 7, pp. 729–747, 2010.
97. J. Y. Chang and S. W. Nam, "Fast Random-Forest-Based Human Pose Estimation Using a Multi-scale and Cascade Approach," *ETRI Journal,* vol. 35, no. 6, pp. 949–959, 2013.
98. D. Wu, C. Jennings, J. Terpenny, R. X. Gao, and S. Kumara, "A comparative study on machine learning algorithms for smart manufacturing: tool wear prediction using random forests," *Journal of Manufacturing Science and Engineering,* vol. 139, no. 7, p. 071018, 2017.
99. K. Tan, H. Wang, L. Chen, Q. Du, P. Du, and C. Pan, "Estimation of the spatial distribution of heavy metal in agricultural soils using airborne hyperspectral imaging and random forest," *Journal of hazardous materials,* vol. 382, p. 120987, 2020.
100. Z. Chunhui, G. Bing, Z. Lejun, and W. Xiaoqing, "Classification of Hyperspectral Imagery based on spectral gradient, SVM and spatial random forest," *Infrared Physics & Technology,* vol. 95, pp. 61–69, 2018.

101. F. Yang, H.-z. Wang, H. Mi, C.-d. Lin, and W.-w. Cai, "Using random forest for reliable classification and cost-sensitive learning for medical diagnosis," *BMC bioinformatics,* vol. 10, pp. 1–14, 2009.
102. E. Elyan and M. M. Gaber, "A fine-grained random forests using class decomposition: an application to medical diagnosis," *Neural computing and applications,* vol. 27, pp. 2279–2288, 2016.
103. D. Yao, J. Yang, and X. Zhan, "A Novel Method for Disease Prediction: Hybrid of Random Forest and Multivariate Adaptive Regression Splines," *J. Comput.,* vol. 8, no. 1, pp. 170–177, 2013.
104. X. Zhang, Z. Shi, X. Liu, and X. Li, "A hybrid feature selection algorithm for classification unbalanced data processsing," in *2018 IEEE international conference on smart internet of things (SmartIoT),* 2018: IEEE, pp. 269–275.
105. J. Sun, C. Li, X.-J. Wu, V. Palade, and W. Fang, "An effective method of weld defect detection and classification based on machine vision," *IEEE Transactions on Industrial Informatics,* vol. 15, no. 12, pp. 6322–6333, 2019.
106. J. Huang *et al.*, "Residual Swin transformer-based weld crack leakage monitoring of pressure pipeline," *Welding in the World,* pp. 1–13, 2023.
107. T. Liu *et al.*, "An expert knowledge-empowered CNN approach for welding radiographic image recognition," *Advanced Engineering Informatics,* vol. 56, p. 101963, 2023.
108. R. Qin *et al.*, "A novel physically interpretable end-to-end network for stress monitoring in laser shock peening," *Computers in Industry,* vol. 155, p. 104060, 2024.
109. F. K. Došilović, M. Brčić, and N. Hlupić, "Explainable artificial intelligence: A survey," in *2018 41st International convention on information and communication technology, electronics and microelectronics (MIPRO),* 2018: IEEE, pp. 0210–0215.
110. C. Molnar, *Interpretable machine learning.* Lulu.com, 2020.

Chapter 2
Optical Spectroscopy-Based Inner Porosity Monitoring for GTAW

2.1 Optical Spectroscopy in Arc Welding

Gas tungsten arc welding (GTAW) stands as a pivotal process within the aerospace industry, a sector known for its stringent demands not only on surface aesthetics but also on internal microstructural integrity and performance. While various non-destructive testing (NDT) methods are employed to scrutinize weld quality, their limitations are evident. Firstly, NDT is typically conducted offline, post-welding, by skilled technicians. Furthermore, it falls short in inspecting every product comprehensively, relying instead on manual sampling inspection. In recent years, as lightweight manufacturing gains precedence and robotic automatic welding advances rapidly, ensuring the high stability and reliability of weld products with zero defects becomes both crucial and complex. This necessitates a paradigm shift towards more robust and efficient inspection methodologies capable of meeting the evolving demands of modern manufacturing.

Real-time monitoring and control of weld quality during dynamic welding processes represent a crucial aspect of ensuring process stability and assessing seam integrity promptly. This approach enables the early detection of abnormalities or inadequacies in welding, offering timely warnings and facilitating immediate corrective actions. By pinpointing weld defects as they occur, it becomes possible to promptly mitigate or manage these flaws throughout the manufacturing process, leading to a substantial optimization of key parameters for enhanced product quality in real-time. Harnessing real-time welding data allows for a more refined enhancement of traditional non-destructive testing (NDT) methods, thereby elevating both their accuracy and efficiency. Furthermore, it empowers intelligent robotic welding systems to operate with greater precision and adaptability. The culmination of these advancements results in welding products characterized by heightened stability and reliability, meeting the exacting standards of modern manufacturing with finesse.

Welding defects are an unavoidable aspect of the intricate dynamics inherent in the welding process, especially notable in Al-Mg alloy gas tungsten arc welding

Z. Zhang et al., *Multiple Sensing-Driven Feature Extraction and Defect Detection*,
https://doi.org/10.1007/978-981-96-4808-5_2

(GTAW). One of the primary culprits behind these defects is the formation of inner porosity, largely attributed to the significant disparity in hydrogen atom solubility between the liquid and solid welding pools. However, a myriad of other factors also contributes to porosity generation, including assembly configuration, welding structure, process parameters, ambient conditions such as temperature and humidity, and the cumulative effect of heat. Despite preemptive measures taken prior to welding, effectively controlling porosity remains a challenge due to its inherent randomness and complexity. Consequently, research into real-time inner porosity detection for Al-Mg alloy in GTAW holds paramount importance in ensuring welding quality. By developing robust methodologies for on-the-fly porosity assessment, manufacturers can better safeguard against defects and uphold the integrity of their welding processes. Sensing and data play crucial roles in the domain of robotic intelligent welding manufacturing. Through the acquisition and real-time processing of pertinent physical information, a welding robot has the potential to elevate its capabilities and assess welding quality seamlessly within dynamic welding operations.

Spectrum emission, as a non-intrusive sensing technology, offers a trove of information, including details about chemical elements. The wavelengths and emission intensities within a spectrum have demonstrated correlations with both in-process seam quality and defects. Throughout the gas tungsten arc welding (GTAW) process, an array of spectrum components corresponding to various chemical elements can be captured, providing extensive insights into welding dynamics. Arc spectroscopy is particularly advantageous due to its ability to furnish rich information in real-time and its relevance to internal metallurgical defects in welding [1, 2]. For instance, Mirapeixa conducted quantitative research on the impact of aluminum (Al) contribution on the seam quality of Usibor blanks in laser welding [3]. They extracted the ratio of line-to-continuum for the Al I emission line at 396.15 nm as a monitoring feature. Song et al. predicted the Al concentration during a laser additive manufacturing process by utilizing the Al/Ti line-intensity ratio and support vector regression [4]. In the realm of defect detection, Zhang proposed several spectrum bands of interest and extracted their statistical features to identify surface oxidation defects [5]. Huang et al. employed a k-medoids algorithm to select spectral lines with heightened sensitivity to seam porosity [6]. Yu et al. extracted the intensity ratio between the H I line (656.28 nm) and Ar I line (641.63 nm) to mitigate interference from welding current pulse and non-hydrogen factors, thereby identifying the most probable position of native porosity [7]. Additionally, Huang et al. used ratios such as H I 656.27 nm/Ar I 675.28 nm and H I 656.27 nm/Ar I 696.57 nm to detect the presence of porosity after eliminating the influence of pulse current using empirical mode decomposition, effectively distinguishing between low and no porosity [8]. These studies underscore the significance of spectrum emission analysis in advancing the understanding and control of welding processes.

In conclusion, significant progress has been made in the research on welding process spectroscopy, yet there are areas that require further development. On one hand, there is a noticeable lack of research reports on the gas tungsten arc welding (GTAW) porosity mechanism, with a dearth of theoretical support for accurate online porosity detection. On the other hand, the methods for mining spectral knowledge are relatively limited, resulting in low information utilization efficiency. There is an urgent need for scientifically effective methods of information mining and the establishment of a strong correlation between porosity and spectral information.

2.2 FastICA-Based Spectral Features Extraction

2.2.1 Experiment Design

Figure 2.1 depicts the intricate architecture of the spectrum acquisition system, which plays a pivotal role in the investigation of welding processes. Throughout the welding operation, both the arc and probe exhibit relative immobility, ensuring consistent data acquisition. The arc light, meticulously captured by the probe, traverses through an optical fiber conduit, ultimately reaching a high-resolution 3648-pixel CCD spectrometer. This state-of-the-art spectrometer, characterized by its remarkable sensitivity, operates within the ultraviolet to near-infrared spectrum, precisely ranging from 200 to 1100 nm. It is pertinent to note that the wavelength resolution of this spectrometer stands at an impressive 0.27 nm, ensuring meticulous data capture. In this experimental setup, the chosen material for investigation is the 5A06 aluminum alloy, renowned for its widespread industrial applications and intriguing metallurgical properties. Detailed welding process parameters are meticulously tabulated in Table 2.1, providing comprehensive insights into the experimental conditions employed.

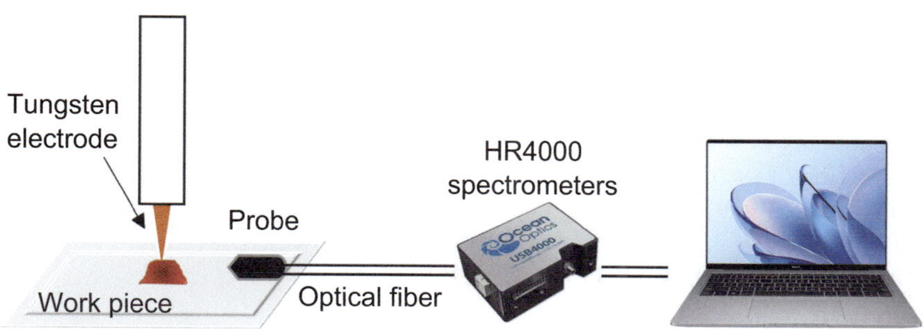

Fig. 2.1 Schematic illustration of spectrum acquisition system

Table 2.1 Welding process parameters

Peak current	Base current	Welding speed	Wire speed	Thickness
230 A	40 A	3 mm/s	7 mm/s	4 mm

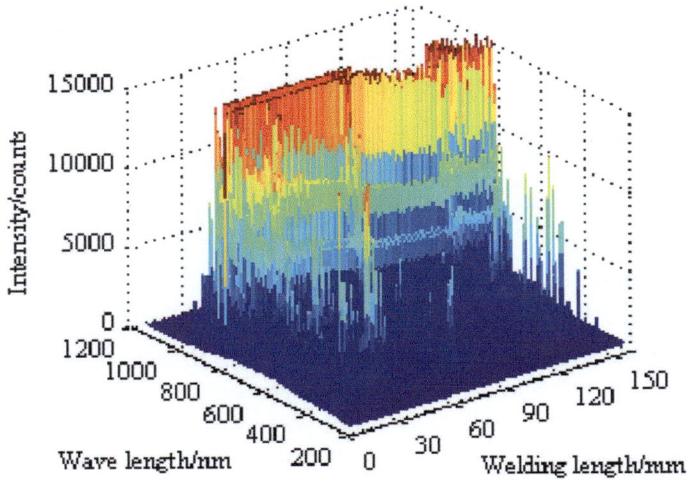

Fig. 2.2 The collected spectral data

Figure 2.2 presents the culmination of the data acquisition process, showcasing the collected spectral data in a visually comprehensible manner. The dataset, characterized by its multidimensional nature, poses significant challenges in direct interpretation and analysis. Specifically, the spectral data spans 3648 dimensions along the wavelength axis and 2400 dimensions concerning the welding length axis. Confronted with such voluminous data, conventional analytical methodologies fall short in efficiently identifying and classifying welding defects.

Thus, in light of the formidable task at hand, the imperative of dimensionality reduction and feature extraction techniques emerges as a pivotal strategy. By systematically distilling the essence of the original dataset, these methodologies aim to unveil underlying patterns and intrinsic characteristics, facilitating a more nuanced understanding of welding defects and their underlying mechanisms. Consequently, the adoption of such sophisticated analytical approaches promises to enrich the efficacy and precision of defect classification and identification within the realm of welding processes.

2.2.2 Spectral Data Analysis Pipeline

The comprehensive data analysis pipeline, delineated in Fig. 2.3, serves as a critical component in elucidating the intricate nuances embedded within the collected spectral data. A systematic approach is undertaken, commencing with the removal of continuous spectra inherent in the raw data acquired by the spectrometer. This initial preprocessing step aims to eliminate extraneous noise and artifacts, thereby enhancing the clarity and fidelity of the spectral information. Subsequently, a judicious selection of spectral lines is meticulously carried out, laying the groundwork for the subsequent application of the Fast Independent Component Analysis (fastICA) technique. This dimensionality reduction method endeavors to distill the salient features encapsulated within the chosen spectral lines, thereby streamlining the computational complexity while preserving the essence of the original dataset. Moreover, the data analysis process encompasses a critical phase devoted to the mitigation of pulse interference, a prevalent source of distortion encountered in spectral measurements. Leveraging advanced signal processing algorithms and filtering techniques, the deleterious effects of pulse interference are effectively attenuated, ensuring the integrity and reliability of the subsequent analytical endeavors. In the final phase of the data analysis endeavor, the focus shifts towards the detection and characterization of porosity defects, a significant concern within the domain of welding inspection. To this end, the Radial Basis Function (RBF) network, renowned for its efficacy in nonlinear classification tasks, is harnessed to discern subtle deviations indicative of porosity defects within the spectral data. Through the judicious orchestration of these analytical methodologies, the data analysis process culminates in the identification and classification of porosity defects, thereby facilitating informed decision-making and quality assurance within the realm of welding processes.

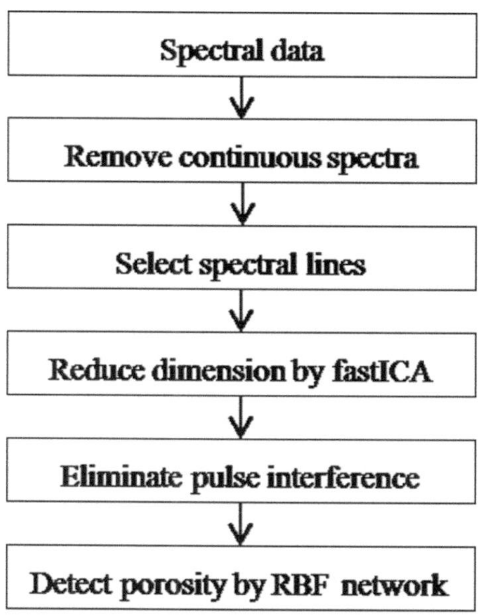

Fig. 2.3 The flow chart of data analysis

2.2.3 *Spectral Data Analysis Process Based on FastICA Method*

The initial phase of the analytical process entails the elimination of continuous spectra, a crucial step in disentangling the complex array of signals inherent within the spectral data. As depicted in Fig. 2.4, the arc spectra captured over a specific duration comprise a combination of continuous spectra and discrete line spectra. The genesis of continuous spectra is multifaceted, stemming from intricate interactions influenced by a myriad of factors, thereby yielding spectral signals characterized by weak radiation intensity. Consequently, continuous spectra are typically disregarded as the primary analytical signals owing to their convoluted nature and limited informational content. Conversely, line spectra emerge as discernible features attributable to transitions between excited states, intricately intertwined with the dynamic interplay of arc behavior and welding parameters. To unveil the underlying line spectra, the initial task necessitates the elimination of continuous spectra. This is achieved through a meticulously orchestrated sequence of computational steps. Initially, the determination of the lower envelope of the original signal is executed, as showcased in Fig. 2.5. However, owing to the inherent limitations stemming from the spectrometer's resolution, a phenomenon of line spectral aliasing ensues, manifesting as the presence of spurious spectral components within the lower envelope curve. In response, a secondary derivation of the lower envelope, referred to as the lower envelope of the lower envelope curve, is meticulously computed to mitigate the effects of spectral aliasing. Subsequent to these iterative envelope calculations, the continuous spectra are successfully isolated, thereby paving the way for the extraction of line spectra through a straightforward subtraction process. The culmination of these analytical maneuvers is exemplified in Fig. 2.6, wherein the outcomes of the secondary envelope calculation are showcased. With the removal of continuous spectra achieved through this iterative process, the pristine line spectra stand delineated, poised for subsequent data processing endeavors.

Fig. 2.4 Arc spectra

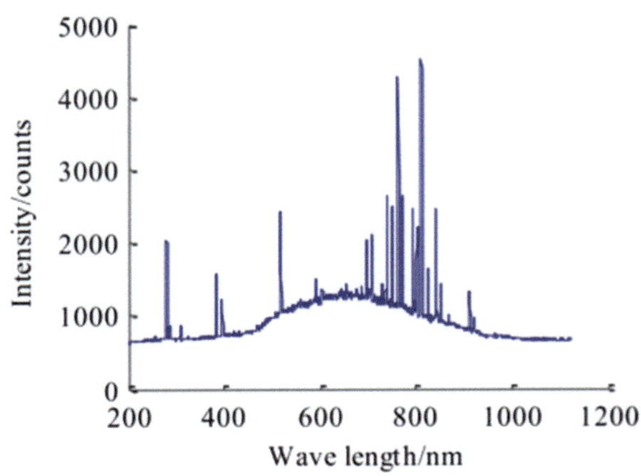

Fig. 2.5 The first lower envelope of arc spectra

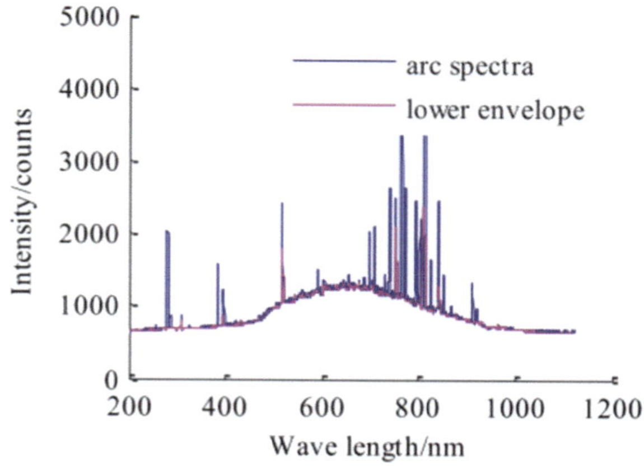

Fig. 2.6 The second lower envelope of arc spectra

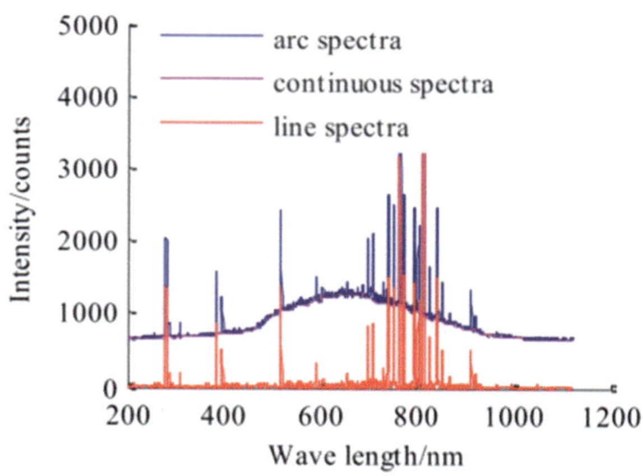

In essence, the meticulous elimination of continuous spectra and the extraction of line spectra underscore the foundational steps in unraveling the intricate spectral signatures inherent within the welding process. Through the judicious application of computational methodologies and signal processing techniques, the analytical framework stands primed to unravel the underlying intricacies embedded within the spectral data, thereby facilitating a more nuanced understanding of welding dynamics and defect characterization.

The subsequent step in the analytical progression involves the judicious selection of spectral lines, a pivotal endeavor aimed at isolating salient features germane to the welding process. As delineated in Fig. 2.6, discernible variations in spectral intensities are evident across distinct wavelength ranges, notably pronounced within the 280–500 and 650–900 nm regions. Notably, the heightened intensities observed within the 280–500 nm range predominantly originate from metallic elements, including aluminum (Al) and magnesium (Mg), whereas spectral emissions within the 650–900 nm range primarily stem from the presence of argon (Ar) element. This delineation underscores the complex interplay of elemental dynamics within the welding environment, each contributing distinctive spectral signatures amenable to analysis and interpretation. Acknowledging the susceptibility of the spectrum acquisition process to extraneous sources of interference, such as ambient light and noise pollution, coupled with the inherent stability of certain spectral lines throughout the welding operation, a pragmatic threshold approach is adopted to filter out low-intensity spectral emissions. Herein, a threshold criterion of 1000 counts is imposed, ensuring the exclusion of spectral signals failing to meet the requisite intensity threshold throughout the entirety of the welding process. Concurrently, leveraging the comprehensive NIST element library facilitates the identification of corresponding elemental compositions associated with the selected spectral lines, thus augmenting the interpretive framework underpinning the analysis.

The culmination of this meticulous selection process culminates in the identification of 22 discernible spectral lines, encompassing a diverse array of elemental contributions. Noteworthy among these are 5 metallic spectral lines, 1 hydrogen (H) spectrum line, and 16 argon (Ar) spectral lines, as showcased in Table 2.2. Importantly, the rationale underpinning the establishment of the 1000 counts threshold is buttressed by findings elucidated within the existing literature. Notably, previous investigations, underscore the sensitivity of argon (Ar) emissions to variations in current and arc length, while highlighting the responsiveness of magnesium (Mg) and hydrogen (H) emissions to fluctuations in argon flow rates. Remarkably, the selected spectral lines encapsulate all elemental sensitivities delineated within the aforementioned literature, thus validating the efficacy of the established threshold criteria in facilitating the discernment of pertinent spectral features germane to welding dynamics.

Subsequent to the selection of 22 spectral lines, it becomes evident that the dataset remains overly voluminous for the precise detection of defects within the welding process. Consequently, the implementation of further processing techniques becomes imperative to mitigate the dimensionality of the dataset, thereby enhancing computational efficiency and facilitating more nuanced defect detection methodologies. To this end, Fast Independent Component Analysis (FastICA) is deployed as a sophisticated dimensionality reduction technique, with the specific objective of distilling the salient features encapsulated within the selected spectral lines. The FastICA algorithm, tailored to operate on the ensemble of 22 selected spectral lines, endeavors to decompose the input matrix X into a set of statistically independent components,

Table 2.2 Some selected spectral lines

Selected wavelength (nm)	NIST wavelength	Chemical species
279.61	280.12	Al II
309.57	309.28	Al I
383.75	383.82	Mg I
656.61	656.28	H I
675.77	675.65	Ar II

thereby affording a succinct representation of the underlying spectral dynamics. Within the algorithmic framework, the number of desired components is meticulously set to 2, with the resultant flow chart elucidated in Fig. 2.7, delineating the intricate sequence of computational operations orchestrated to achieve dimensionality reduction.

Concomitantly, the input matrix X, comprising the ensemble of selected spectral lines, is subjected to meticulous scrutiny, with M denoting the total number of spectral lines under consideration, herein established as 22. The ensuing outcome, as depicted in Fig. 2.8 (blue line), reveals discernible fluctuations characterized by the initial instability of Independent Component 1 (IC1) and Independent Component 2 (IC2), attributable to transient arc dynamics. Moreover, conspicuous fluctuations manifest at the seam position, indicative of underlying porosity defects within the

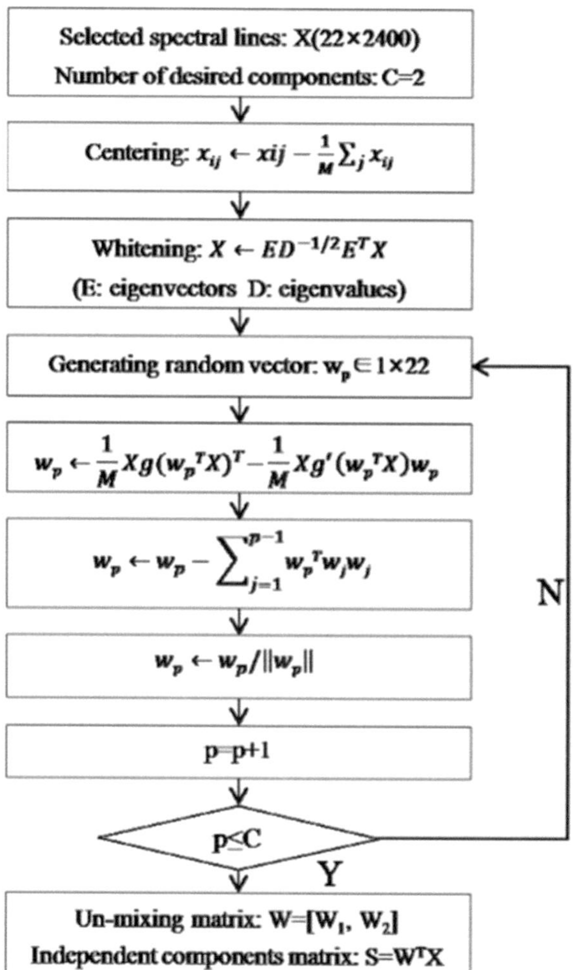

Fig. 2.7 Flow chart of fastICA

welding structure. This discernment underscores the pivotal role assumed by IC1 and IC2 as characteristic signals amenable to defect detection, with IC1 particularly exhibiting a pronounced correlation with the manifestation of porosity defects. The elucidation of these characteristic signals heralds a pivotal breakthrough in the realm of defect detection within welding processes, offering unprecedented insights into the underlying structural integrity and facilitating informed decision-making pertaining to quality assurance protocols.

In the ultimate phase of the analytical endeavor, the focus converges on the eradication of pulse interference, an omnipresent challenge confounding defect detection methodology within welding processes. As delineated in Fig. 2.8, the discernment of periodic fluctuations exhibited by Independent Component 1 (IC1) and Independent Component 2 (IC2) underscores the deleterious impact of pulse interference on the

Fig. 2.8 Independent components extracted by FastICA

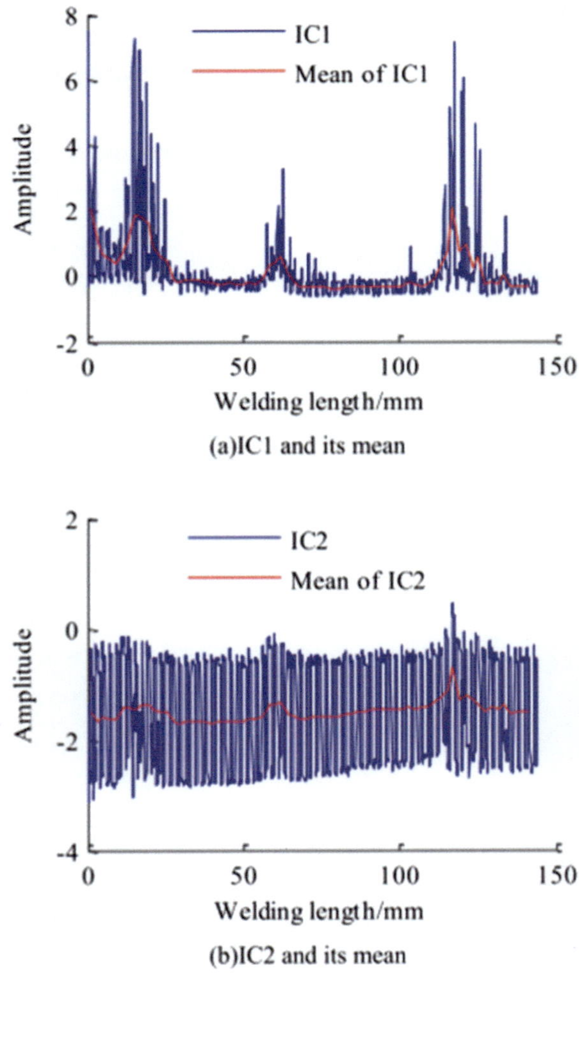

fidelity and interpretability of the spectral signals. This perturbation, stemming from transient variations in welding arc dynamics, introduces spurious oscillations that obfuscate the underlying defect signatures, thereby impeding the efficacy of defect detection methodologies. In response to this formidable challenge, a strategic intervention is devised to mitigate the deleterious effects of pulse interference, thereby restoring clarity and coherence to the spectral signals. This intervention entails the substitution of the original IC1 and IC2 signals with their respective mean values computed over the signal's pulse period. As exemplified in Fig. 2.8, this transformative adjustment engenders a perceptible refinement in signal stability, with the resultant mean value signals evincing a discernible attenuation of pulse-induced oscillations.

Indeed, the salutary effects of this modification are readily discernible, as the erstwhile erratic fluctuations give way to a smoother and more discernible mean value signal. This perceptible enhancement in signal coherence augurs well for defect detection endeavors, as the resultant signals are endowed with greater discernibility and interpretability vis-à-vis their pulse-interference-laden counterparts. Consequently, the adoption of this strategic intervention marks a significant milestone in the quest for precision defect detection within welding processes, heralding a transformative advancement in the analytical framework underpinning quality.

2.2.4 Detect Porosity by RBF Network

This study adopts an innovative approach, leveraging the Radial Basis Function (RBF) network for the precise detection of porosity defects within welding processes. Specifically, the mean value signals of Independent Component 1 (IC1) and Independent Component 2 (IC2) serve as inputs to the RBF network, thereby facilitating a comprehensive analysis of spectral dynamics indicative of underlying defects. To rigorously evaluate the efficacy of the proposed methodology, a meticulous selection of training samples is undertaken, sourced from the 11–41 mm section of the welding workpiece. Within this delineated range, samples spanning the 11–26 mm segment represent instances of porosity defects, while the 26–41 mm range corresponds to correctly welded sections. Subsequently, test samples are drawn from the entire welding process, encompassing a holistic spectrum of spectral variations reflective of diverse welding conditions and defect manifestations.

The outputs of the RBF network, as depicted in Fig. 2.9a, offer compelling insights into the spatial distribution of porosity defects, with discernible anomalies evident within the approximate regions of 56–70 mm and 105–135 mm. Importantly, these detected defects align with the physical condition of the welding workpiece, as corroborated by the corresponding image showcased in Fig. 2.9b. This tangible correlation underscores the efficacy of the RBF network in delineating and characterizing porosity defects, thereby facilitating informed decision-making and quality assurance protocols within welding processes.

Fig. 2.9 Result of the first experiment

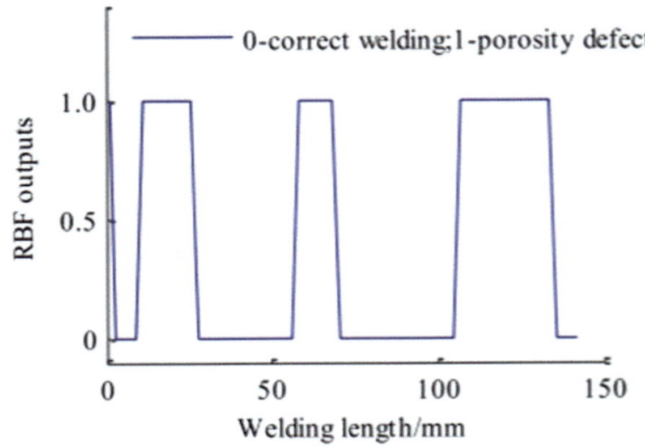

(a)RBF outputs of the first experiment

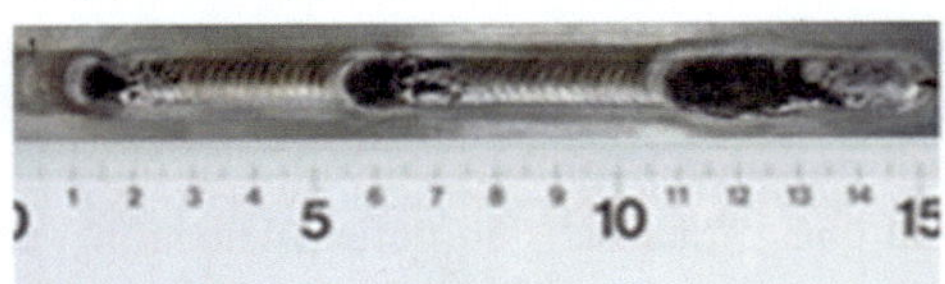

(b)Picture of the first welding workpiece

In a comparative analysis, the outcomes derived from Principal Component Analysis (PCA) in conjunction with the RBF network are juxtaposed against those obtained through the proposed methodology. As elucidated in Fig. 2.10, it becomes apparent that PCA fails to accurately detect the porosity defect within the approximate 56–70 mm range, underscoring the superior performance of the proposed methodology vis-à-vis PCA. Notably, this comparative assessment highlights the inherent advantages conferred by the proposed methodology, thereby affirming its efficacy in defect detection and characterization endeavors.

Fig. 2.10 Result of using PCA-RBF of the first experiment

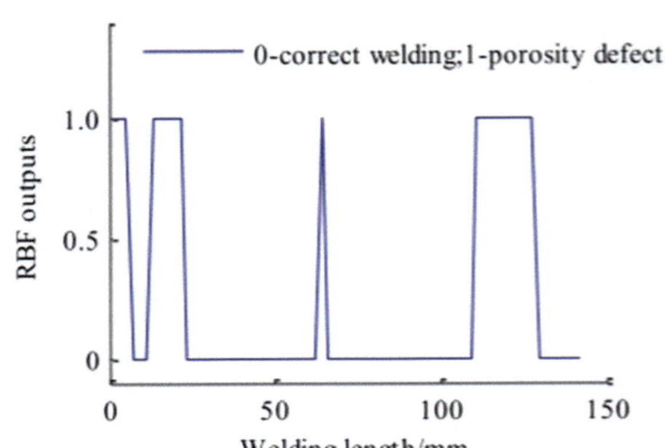

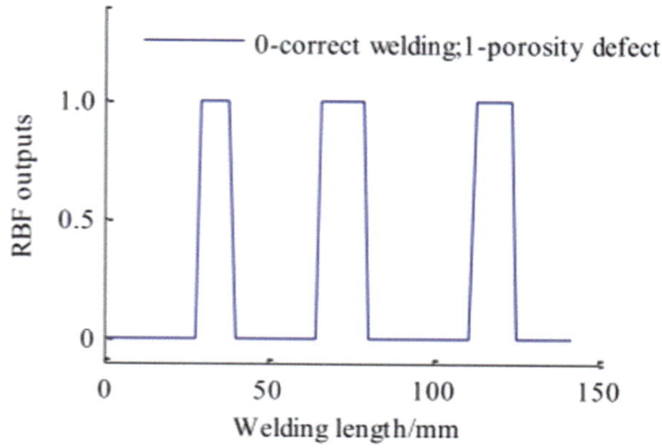

(a)RBF outputs of the second experiment

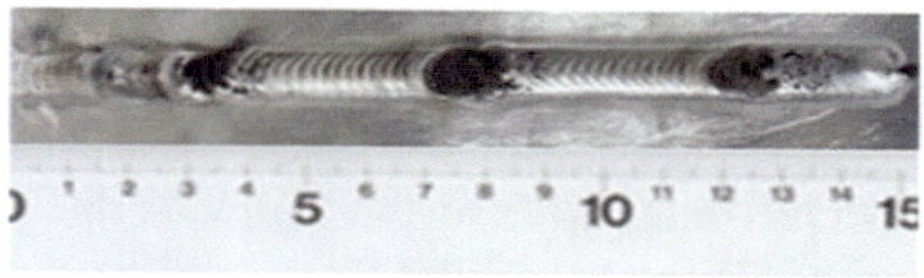

(b)Picture of the second welding workpiece

Fig. 2.11 Result of the second experiment

Building upon these foundational insights, a secondary experimental iteration is conducted, wherein training samples are extracted from the signals spanning the 30-50 mm section of the welding workpiece. Within this designated range, the 30–40 mm segment represents instances of porosity defects, while the 40–50 mm range comprises samples indicative of correctly welded sections. Subsequent analysis, as depicted in Fig. 2.11a, reveals the successful detection of porosity defects within the regions of approximately 70–80 and 120–130 mm, duly excluding the 30–40 mm section earmarked for training purposes. Once again, a salient correlation emerges between the detected defects and the physical state of the welding workpiece, as evidenced by the corresponding image showcased in Fig. 2.11b.

Furthermore, Fig. 2.12 offers a comparative assessment of the outcomes derived from PCA and the RBF network, reaffirming the superior performance of the proposed methodology in effectively detecting all porosity defects. This empirical validation underscores the inherent advantages conferred by the proposed methodology, positioning it as a potent tool for defect detection and characterization within welding processes. In summation, the proposed methodology exhibits a marginal yet discernible advantage over PCA, thereby underscoring its efficacy and relevance in advancing the frontier of defect detection methodologies within welding processes.

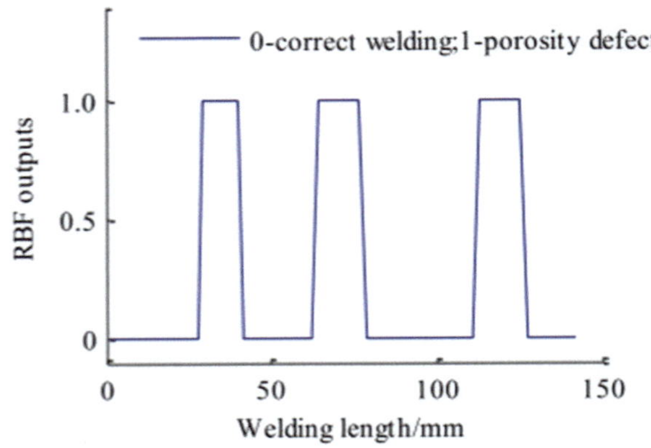

Fig. 2.12 Result of using PCA of the second experiment

2.2.5 Conclusion

This paper presents a novel methodology for the detection of porosity defects inherent in pulsed TIG (Tungsten Inert Gas) welding processes, employing a sophisticated amalgamation of Fast Independent Component Analysis (fastICA) and Radial Basis Function (RBF) network algorithms. The devised approach encompasses a multi-step process aimed at meticulously parsing spectral data, isolating salient features, and discerning characteristic defect signatures within the welding environment.

The methodology commences with the meticulous removal of continuous spectra through iterative lower envelope calculations, thereby attenuating extraneous noise and enhancing the signal-to-noise ratio inherent in the acquired spectral data. Subsequently, a judicious selection of 22 spectral lines is undertaken, predicated upon a predefined threshold criterion, and juxtaposed against the comprehensive NIST element library to ascertain elemental compositions. This discriminative process ensures the inclusion of spectral lines most indicative of underlying welding dynamics and defect manifestations.

Building upon this foundation, fastICA is deployed to extract two independent components from the selected spectral lines, thereby distilling the essence of the spectral data while mitigating the deleterious effects of pulse interference. This strategic intervention, facilitated by the computation of mean values over the pulse period, serves to enhance signal stability and coherence, thereby fortifying the analytical framework underpinning defect detection endeavors.

The culminating phase of the proposed methodology involves the deployment of the RBF network for the precise detection of porosity defects within the welding structure. Experimental validation underscores the efficacy and reliability of the proposed methodology, affirming its capacity to accurately discern and characterize porosity defects within aluminum alloy pulse TIG welding processes. Importantly, the demonstrated accuracy and reliability of the proposed methodology offer compelling evidence of its potential as a pivotal tool for enhancing the efficiency of welding defect detection and contributing to the overarching goal of advancing welding automation processes.

In summation, the proposed methodology represents a significant advancement in the realm of welding defect detection, offering a potent blend of computational algorithms and analytical techniques tailored to the intricacies of pulsed TIG welding processes. By offering a robust framework for discerning and characterizing porosity defects, the proposed methodology stands poised to catalyze transformative advancements in welding quality assurance protocols, thereby bolstering the efficacy and reliability of industrial welding operations.

2.3 PCA and t-SNE-Based Spectral Features Extraction

2.3.1 Experiment Setup and Testing

The experimental setup comprised a sophisticated amalgamation of a multisensory signal acquisition apparatus and a robotic arc welding and process control system. Illustrated in Fig. 2.13a, the multisensory signal-acquisition system gathered data through arc spectroscopy, welding pool imaging, arc sound pressure, arc current, and arc voltage. Specifically, the arc spectroscopy signal was procured via a CCD spectrometer (AvaSpec-1350F-USB2) equipped with an optical probe, fiber, and spectrometer, capable of capturing wavelengths ranging from 350 to 1100 nm, boasting 1,350 pixels, and offering a minimum integration time of 0.7 ms. Figure 2.13b depicts the pre-welding experimental alignment of the spectrum probe. The audible sound sensing component featured an omnidirectional capacitance microphone coupled with a signal conditioner, strategically positioned at a 75° angle above the workpiece. Meanwhile, the welding arc current and arc voltage signals were meticulously obtained using a Hall-effect sensor operating at a sampling rate of 40 kHz. A robotic vision system outfitted with a CCD camera (MER-200-14GM) was intricately devised to capture high-resolution images of the welding pool. For gas shielding during the robotic welding procedure, pure Ar (99.99%) was employed. The welding process and its speed were deftly controlled by a Yaskawa Industry robot. Moreover, an adeptly crafted LabVIEW application software was developed to synchronize the triggering, acquisition, and storage of all signals simultaneously, while also providing control over welding current, wire feed speed, data collection duration, and frequency.

Table 2.3 provides an exhaustive breakdown of the welding parameters pertinent to Al-Mg alloy (5A06) during pulsed gas tungsten arc welding (GTAW). The deliberate introduction of an oxidation film, recognized as a pivotal contributor to porosity occurrence, was meticulously orchestrated to induce inner porosity within the weld. In order to comprehensively assess the micro-characterization of this porosity, a rigorous regimen of post-non-destructive testing (NDT) and subsequent destructive testing was meticulously executed. Intriguingly, the investigation delved deeper into understanding the intricate distribution patterns of porosity within the welded seams.

Fig. 2.13 Experimental system for robotic GTAW

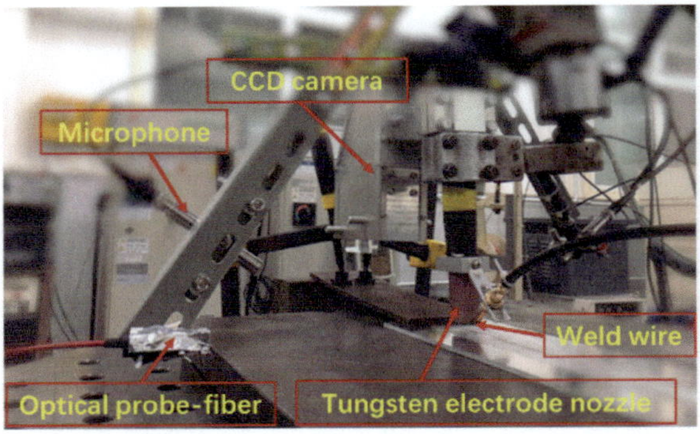

(a) Multisensory signal-collection equipment.

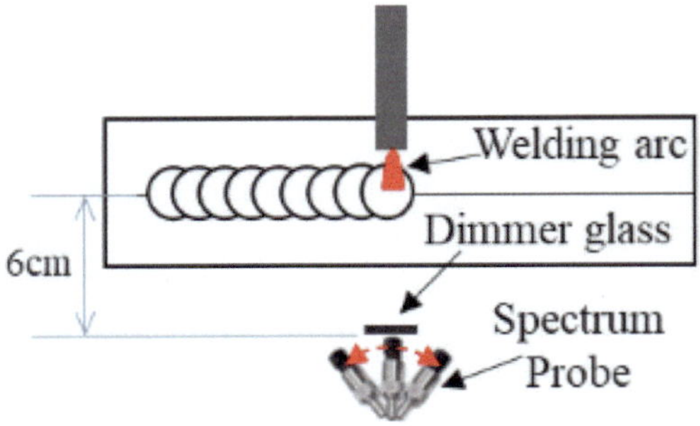

(b) Adjusted position of spectrum probe.

To unravel the enigmatic rules governing porosity distribution, both transverse and longitudinal sections of the weld were meticulously prepared and scrutinized under an optical microscope. This meticulous observation protocol closely followed the established methodology outlined in literature, ensuring consistency and comparability of results across studies.

Through a deliberate manipulation of the surface oxidation film, a diverse array of welding seams exhibiting varying degrees of porosity were meticulously engineered, as vividly illustrated in Fig. 2.14 This intricate process involved subjecting the plate

Table 2.3 Parameters for Al-Mg alloy butt-welding

Parameter	Value	Parameter	Value
Welding peak current (A)	240	Welding speed (cm/min)	16
Welding base current (A)	50	Electrode diameter (mm)	3.2
Pulse duty ratio (%)	50	Wire diameter (mm)	1.6
Ar flow (L/min)	15	Wire feed speed (mm/s)	12
Pulse frequency (Hz)	1	Plate size (mm)	300 × 50 × 4

to eight spot welding pulses without the provision of preheating, thus initiating selective removal of the oxidation film. The consequential welding seams, showcased in Fig. 2.15, offer a multifaceted perspective comprising front and back views alongside X-ray detection results, each unveiling unique porosity characteristics. Notably, Seam #4, characterized by a meticulous removal of the oxidation film from its first half, displayed a moderate level of porosity, discernibly highlighted by a conspicuous red ellipse. Conversely, Seam #6 exhibited minimal porosity, indicative of a refined welding process. However, the most striking revelation was witnessed in Seam #9, where the oxidation film remained intact, resulting in a cascade of severe porosity reminiscent of a chain formation, prominently encircled by a red ellipse. Notably, the surface porosity adorning the edges of Seam #9 is meticulously delineated in the enlarged inset F, providing a closer examination of this intriguing phenomenon.

Figure 2.16 unveils a comprehensive examination of both transverse and longitudinal sections of porosity-laden Seam #4, wherein the oxidation film remained

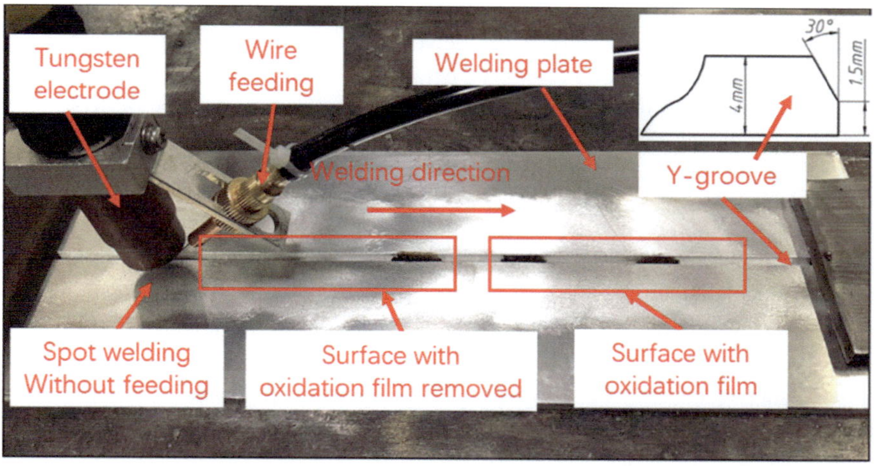

Fig. 2.14 Experiment design for porosity

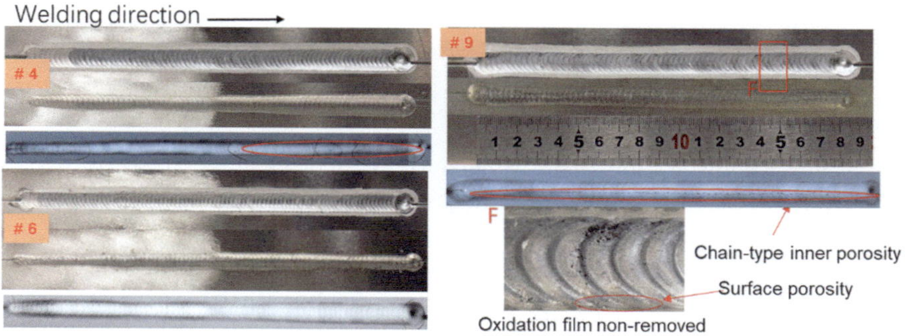

Fig. 2.15 X-ray test results for welding seams with different levels of porosity

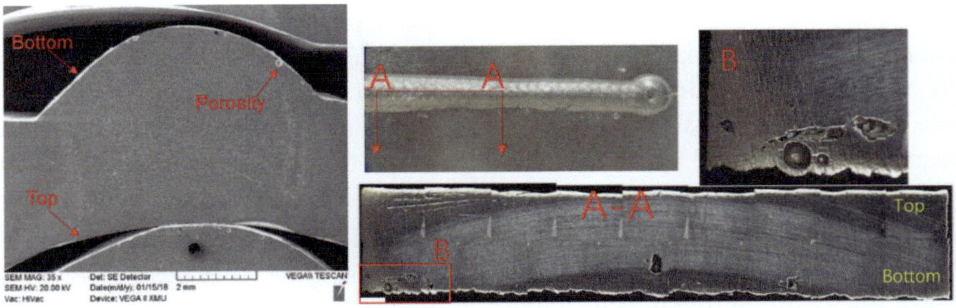

(a) Transverse section. (b) Longitudinal section

Fig. 2.16 Porosity distribution in seam section

intact throughout the welding process. Figure 2.16a provides a detailed insight into the distribution pattern of porosity, notably concentrated along the top and bottom of the seam, with some clusters observed near the fusion line, as discernible from the illustration. Meanwhile, Fig. 2.16b delves into the longitudinal section (A-A) of the welding seam, unraveling a nuanced perspective wherein varied sizes of porosity have manifested predominantly along the seam's bottom, mirroring the transverse section's patterning depicted in Fig. 2.16a. However, amidst the intricate porosity formations, several irregularly shaped defects have also been identified, indicative of potential instances of inadequate fusion. The genesis of these porosity phenomena can largely be attributed to the differential dissolution of hydrogen (H) within the liquid and solid phases of the aluminum alloy welding pool. Typically, hydrogen is sourced from multiple origins including the refractory oxidation film, shielding gas, and feeding wire. From a nucleation standpoint, the refractory oxidation film emerges as a pivotal factor, potentially fostering conditions conducive for the initiation and subsequent growth of porosity. Furthermore, the presence of porosity along the bottom of the seam may be ascribed to inadequate shielding of the seam's rear portion. Moreover, the accelerated solidification rate within the weld pool may have impeded the escape of porosity, exacerbating its formation.

2.3.2 Preprocessing and PCA Analysis of Arc Spectral Signals

Within this section, a comprehensive analysis of arc optical spectroscopy signals is undertaken, encompassing critical stages such as preprocessing, line spectrum selection, and feature reduction.

Figure 2.17 presents an extensive array of multisensory signals captured concurrently during the robotic pulsed gas tungsten arc welding (GTAW) process applied to aluminum alloy. These signals encompass a diverse range of modalities, including welding arc optical spectrum, welding pool imagery, audible sound, arc voltage, and welding current, collectively providing a holistic perspective on the welding process dynamics. Of particular interest is the weld arc optical spectrum signal, meticulously delineated in the upper inset of Fig. 2.5. This signal comprises both a line spectrum and a background spectrum, each offering unique insights into the composition and behavior of the welding arc. The line spectrum predominantly emanates from elemental constituents such as the argon (Ar) shielding gas, the metal base material, and the feeding wire, each contributing distinct spectral signatures. In contrast, the background spectrum typically arises from phenomena within the liquid welding pool and interactions involving the tungsten head, offering complementary information regarding the arc dynamics. A notable observation from the depicted arc spectrum signal is the discernible variation in spectrum intensities corresponding to different phases of the welding process, notably the welding peak current and base current. This nuanced variation underscores the dynamic nature of the welding arc, reflecting fluctuations in temperature, material composition, and arc stability throughout the welding operation. By amalgamating these diverse multisensory signals, this comprehensive dataset facilitates a deeper understanding of the intricate interplay between process parameters and welding outcomes, paving the way for enhanced control strategies and optimization techniques in robotic pulsed GTAW applications for aluminum alloys.

In the pursuit of precision in extracting the emission intensity of a line spectrum within arc welding, it becomes imperative to eliminate the influence of the background spectrum. Figure 2.18 delineates the extraction process of a line spectrum from the original spectrum, achieved through a two-time lower enveloping technique. In the initial step, as illustrated in Fig. 2.18a, the original spectrum undergoes the process of first lower enveloping. This operation effectively delineates the overarching spectral profile, while certain distinctive peaks, denoted by black rectangles, persist amidst the enveloped background. Subsequently, in Fig. 2.18b, the outcome of the second lower enveloping operation is showcased, wherein the background spectrum becomes distinctly separated from the spectral lines of interest. This successful separation renders the extracted spectrum primed for subsequent processing, notably for principal component analysis (PCA) and further analytical endeavors. Such refined extraction methodologies are paramount for enhancing the accuracy and reliability of spectroscopic analysis in welding applications, thereby contributing to advancements in process monitoring and control strategies.

Upon extraction, the line spectrum emerges as a vast ensemble comprising thousands of individual components, necessitating a process of reduction to identify and isolate spectral lines most pertinent to the phenomena of inner porosity and the dynamic nature of the welding process. In addressing this challenge, principal component analysis (PCA) emerges as a powerful linear dimensionality reduction technique within the realms of machine learning and artificial intelligence. PCA operates by distilling the abundance of spectral information into a more manageable subset, thereby facilitating the identification of key features that encapsulate the underlying dynamics of the welding process. By assigning differential coefficients to these features, PCA effectively constructs new dimensions that capture the

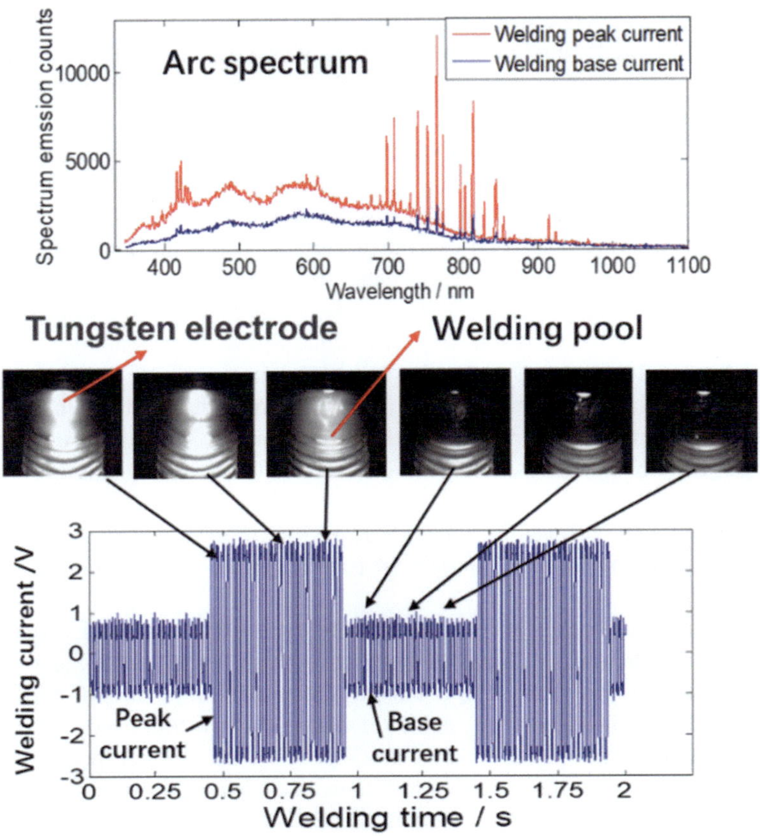

Fig. 2.17 Welding arc spectrum, pool image, and arc current signals simultaneously acquired during Al alloy pulsed GTAW process

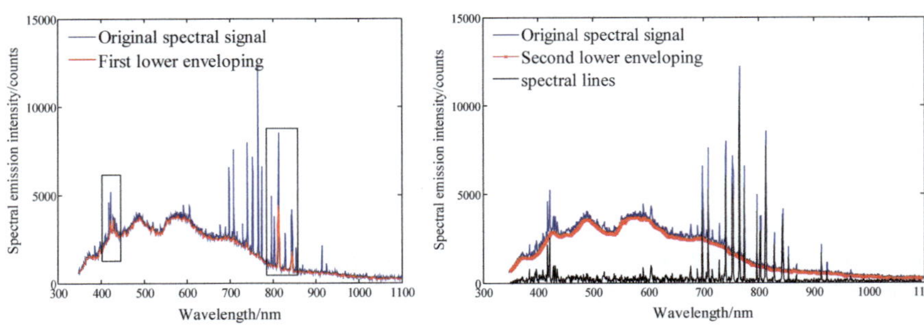

(a) First lower enveloping. (b) Second lower enveloping.

Fig. 2.18 Separating background and line spectrum using two-time lower enveloping

essence of the original data while mitigating redundancy and noise. This transformative process not only streamlines the dataset but also enhances interpretability by accentuating the most salient aspects of the spectral information. By harnessing the capabilities of PCA, researchers can unlock invaluable insights into the intricate interplay between spectral features and welding phenomena, thereby paving the way for enhanced understanding and control of the welding process.

Step 1. Preparing D-dimensional vector representation of N samples for different conditions of welding seam, *X(N,D)*, where D is the dimension of spectrum signal, e.g., 1322; each one and depends on the length of welding seam.
Step 2. Calculating the eigenstructure decomposition

$$\lambda_i e_i = Qe_i, \quad Q = XX^T \tag{2.1}$$

where λ_i is the eigenvalue associated with the eigenvector e_i,
Step 3. The new feature vectors Y are defined by

$$Y_j = W^T x_i; \ (j : 1; \ldots; P; i : 1, \ldots 1322) \tag{2.2}$$

where W is the linear transformation in which the original feature space with D-dimensional is mapped onto a feature subspace with p-dimensional ($D > p$). The columns of W are the eigenvalues e_i in descending order.

In this study, principal component analysis (PCA) was harnessed to effectively reduce the dimensionality of both the background spectrum and the line spectrum, enabling a quantitative assessment of spectral line sensitivity to porosity defects. Eigenvalues, denoted as "ei," played a pivotal role in shaping the new principal components (PCs), serving as a quantitative index to gauge the sensitivity of spectral lines to seam defects.

Figure 2.19 provides a comprehensive depiction of the outcomes yielded by PCA applied to the background spectrum signal, as extracted in Fig. 2.6b. Here, the input dataset denoted as X (3061,1322), encompassing the entirety of the welding seam (approximately 240 mm) and encompassing eight pulse cycles of preheating via spot welding. Figure 2.7a portrays the coefficient curves of PC1, PC2, and PC3, where PC1 (represented by the black line) exhibits a trend akin to that of the original background spectrum in Fig. 2.6, characterized by a positive coefficient value. Notably, the magnitude of the coefficient value directly correlates with the sensitivity of the respective component to the welding process. By establishing a threshold line with an absolute value of 0.03, the three PCs delineated in Fig. 2.19a effectively extracted wavelength bands featuring differing coefficient values. Subsequently, Fig. 2.19b showcases the newly constructed PCs, which display varying intensities due to their unique linear combinations of coefficient values. It is evident that PC1 exhibits a notably higher

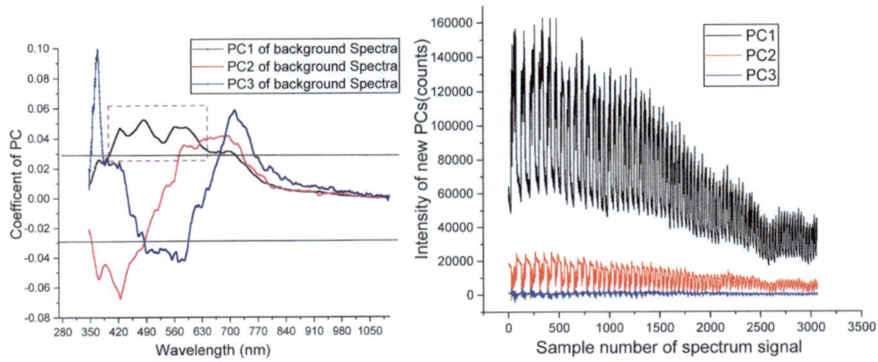

(a) Coefficient curves for top three PCs.

(b) Top three PCs.

Fig. 2.19 PCA of background spectrum signal

intensity value, accompanied by a discernible descending trend. Further analysis reveals that the heightened coefficient value of PC1 within the wavelength range of 400.03–649.78 nm implicates the blackbody emission of the background spectrum, encompassing elements such as the welding pool, tungsten electrode, and feeding wire. Through this rigorous analytical approach, the sensitivity of spectral lines to seam defects is effectively delineated, paving the way for enhanced understanding and control of the welding process.

PCA was applied to the line spectrum extracted from Fig. 2.17b, resulting in insightful observations depicted in Fig. 2.20a. Notably, the line spectrum with higher coefficient values, encompassing the Ar atom spectrum (Ar I) of PC1 and the metal spectrum of PC2 and PC3, is prominently delineated, while the H atom spectrum (H I) exhibits relatively smaller values. These distinct chemical components comprising each selected line spectrum were meticulously identified utilizing the National Institute of Standards and Technology (NIST) database and are comprehensively listed in Table 2.2. The corresponding new PC curves showcased in Fig. 2.20b unveil varying change patterns, each offering unique insights into the dynamic welding process. For instance, the first new feature curve of PC1 exhibits minimal change in the upper envelope, attributable to the modest base welding current of 50 A, which emits minimal arc light while manifesting increased blackbody emission. Furthermore, the descending trend observed, akin to Fig. 2.20b, can be attributed to the negative value of the Ar spectrum lines within PC1. Additionally, during the initial eight pulse cycles, which were executed via spot welding without feeding wire, PC3 demonstrates subtle peaks, contrasting with the heightened intensity observed in PC2 compared to subsequent welding pulse cycles. The application of PCA to both background spectrum (BS) and line spectrum (LS) proves highly effective in mitigating disturbance from background emission and extracting diverse aspects of spectrum information pertaining to seam defects.

To further elucidate the overarching trends governing line spectrum dynamics in response to the dynamic welding process, additional welding seams were repetitively

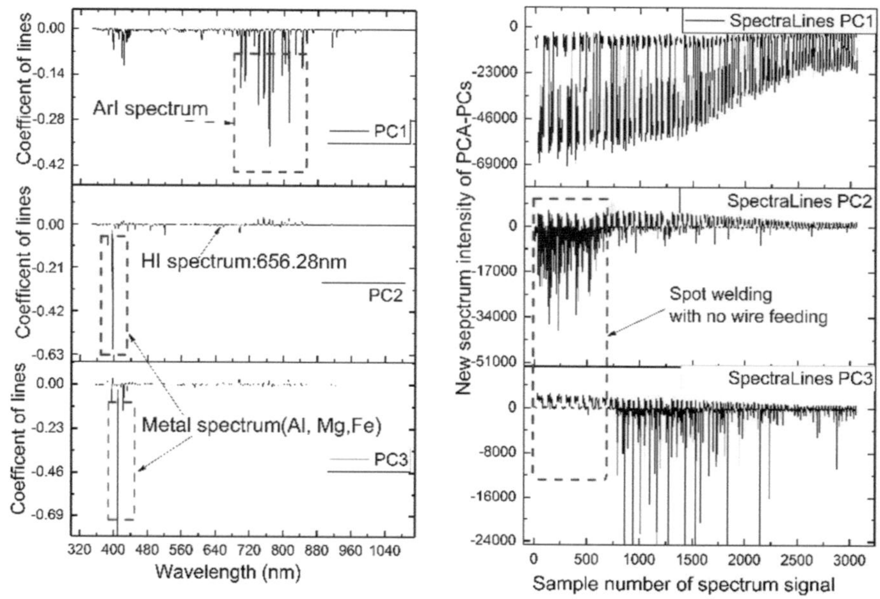

(a) Coefficient of PCs.

(b) Top three PCs.

Fig. 2.20 PCA of line spectrum signals

executed with PCA. This rigorous analysis aimed to ascertain general rules governing line spectrum behavior. Consequently, several key spectral lines pertinent to Al alloy in pulsed GTAW with a wire filler were identified based on PC coefficient values and corroborated by the NIST database, as detailed in Table 2.4. However, it is noteworthy that due to the CCD-pixel limitation of the spectrometer, precise identification of chemical elements for certain spectral lines such as Fe I, Al I, and Mg I posed challenges. Despite this limitation, the insights gleaned from this comprehensive analysis are invaluable for enhancing understanding of the complex manufacturing process and enabling effective detection of seam defects. Future endeavors may entail the augmentation of pixel count to achieve more precise line identification, further enriching our understanding of welding dynamics.

2.3.3 *Spectrum Feature Extraction and Reduction Based on PCA and t-SNE*

This section delves into the extraction of features from optical spectra for the purpose of online detection of varying degrees of inner porosity. Additionally, the study investigates the selection of spectral lines strongly associated with inner porosity through the utilization of PCA and its corresponding coefficients. The correlation and

Table 2.4 General spectral lines selected by PCA

Chemical element	Number of pixels	Observed wavelength (nm)
Ar I	563	675.3
	600	696.54
	618	706.72
	674	739.77
	718	763.51
	802	811.53
	857	842.46
Fe I/Al I/Mg I	105	408.93
	128	422.55
	81	394.69
Mg I	64	384.6
Al I	85	397.07
H I	530	657.27

effectiveness of these selected features are rigorously assessed through quantitative evaluation utilizing statistical metrics.

The practice of utilizing the intensity ratio of hydrogen (H) to argon (Ar) spectral lines is commonplace in porosity detection. Nonetheless, when considering the myriad of Ar spectral lines available, quantitatively selecting a suitable Ar line for ratio calculation poses a significant challenge. In this investigation, the PC coefficient emerged as a pivotal criterion for selection, enabling the identification of two Ar spectral lines with contrasting PC1 coefficients for comparison and subsequent feature construction. As depicted in Fig. 2.21, the spectrum intensity ratio between H I (656.28 nm) and Ar I (763.3 nm) for welding Seam #4 exhibited minimal noise and a distinct change trend, while the ratio between H I and Ar I (675.3 nm) displayed more pronounced noise. Consequently, the Ar I spectral line with a higher PCA coefficient, such as – 0.4182, proved notably more effective than its counterpart at – 0.019 in capturing dynamic changes within the welding process. In essence, the PCA coefficient values derived from the first principal component serve as a potent data mining tool, facilitating the quantitative selection of spectral lines directly associated with seam defects and process variability. This innovative approach underscores the utility of PCA coefficients in enhancing the efficacy of spectral line selection, thereby empowering researchers to discern and quantify critical features essential for porosity detection and process optimization in welding applications.

Figure 2.22 exhibits three distinct process monitoring curves derived from the utilization of the selected Ar I (763.3 nm) spectral line and its corresponding intensity ratio across various degrees of seam porosity. Notably, the presence of pulse welding current introduces interference, markedly evident within the curves. Despite the discernible impact of pulse interference, distinguishing between the three welding processes based solely on porosity levels proves challenging (Fig. 2.23).

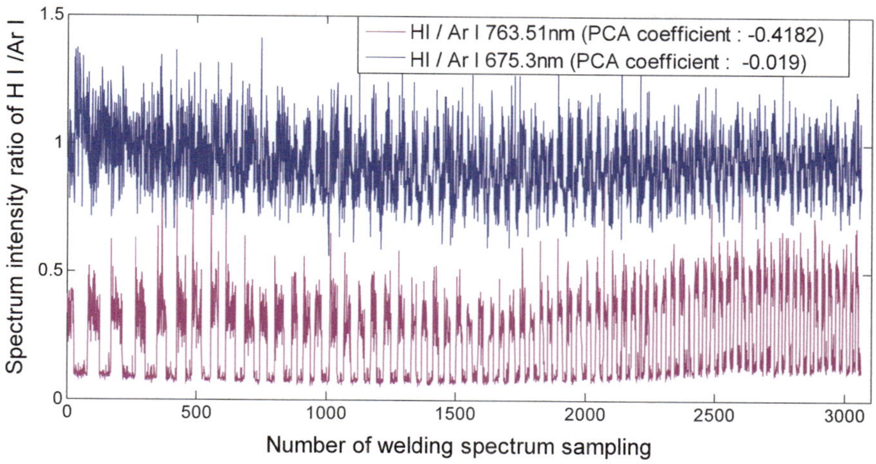

Fig. 2.21 Spectrum intensity ratio between H I (656.28 nm) and Ar I (763.3 and 675.3 nm) with different coefficients

The pulse interference evident in Fig. 2.22 stemmed from the utilization of pulsed welding current. To mitigate this interference, a new monitoring parameter was introduced by calculating the variance of 30 spectrum feature samples, as depicted in Fig. 2.22 However, in Fig. 2.11, displaying monitoring curves for three distinct levels of porosity seams using the intensity ratio of H/Ar, it becomes apparent that distinguishing between these levels remains challenging. Notably, even severe porosity observed in Seam #9 is indistinguishable from Seam #6, which exhibits no porosity.

The novel principal components (PCs) derived from line spectra via PCA amalgamate information from various spectral lines, such as H I, Ar I, Mg I, Fe I, among others, each weighted differently. Consequently, these PCs hold promise as effective feature parameters for monitoring porosity in Al alloy welding processes. Table 2.5 enumerates the pertinent variables to consider when formulating new parameters. Of particular significance for different PCs is the coefficient value associated with the H I spectrum, which significantly influences the resultant feature value. Notably, a higher coefficient value correlates with a more accurate reflection of dynamic changes in the H atom. Additionally, the statistical features outlined in Table 2.5—including variance, kurtosis, and skewness—of specific PC samples, offer substantial potential in mitigating pulse noise and are ripe for extraction for real-time porosity monitoring applications. Through an iterative process involving the adjustment and comparison of different sampling numbers—such as 10, 30, and 60—it was determined that the variance derived from 30 samples emerged as the optimal feature parameter for detecting inner porosity. This meticulous approach ensures the selection of an effective and robust feature parameter, thus enhancing the accuracy and reliability of porosity detection in Al alloy welding processes.

Therefore, based on the selected sample number of 30 and statistical feature of variance, the feature parameter, VarPCLS (variance of PCs of line spectrum) was

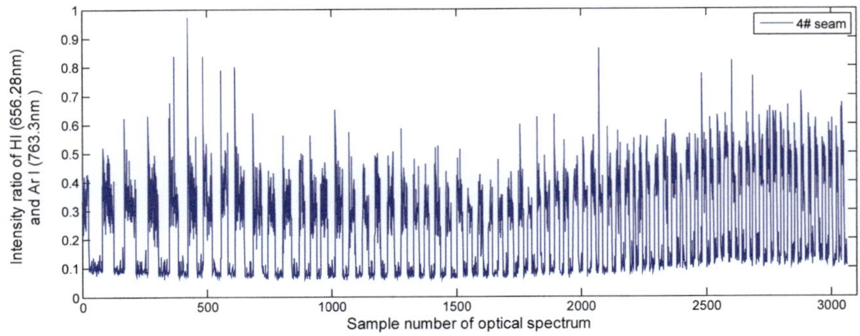

(a) Feature (H I/Ar I) curve for welding Seam #4 with moderate porosity.

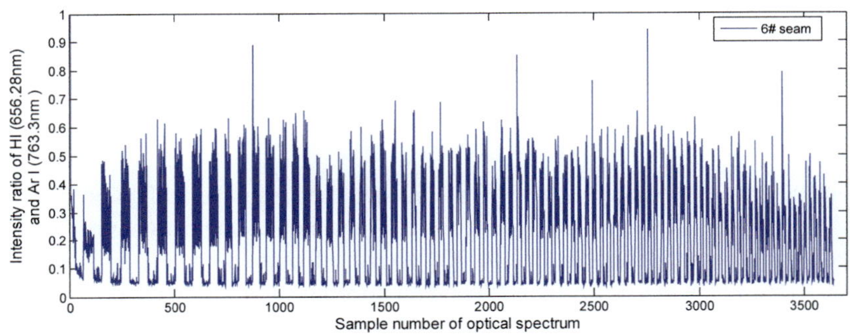

(b) Feature (H I/Ar I) curve for welding Seam #6 with no porosity.

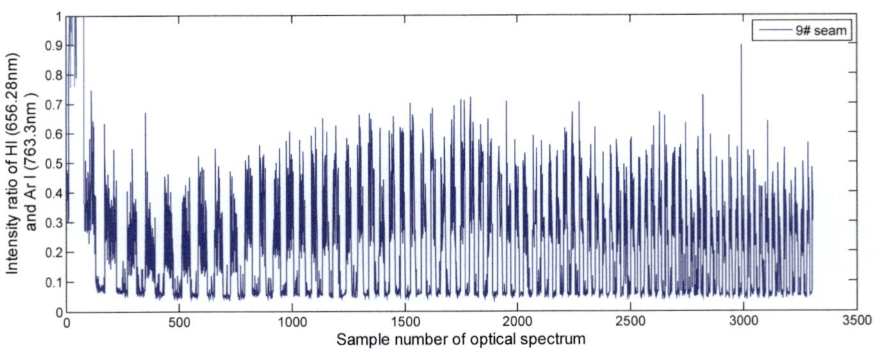

(c) Feature (H I/Ar I) curve for welding Seam #9 with severe porosity.

Fig. 2.22 Spectrum intensity ratio (H I (656.28 nm)/Ar I (763.3 nm)) for different welding seams

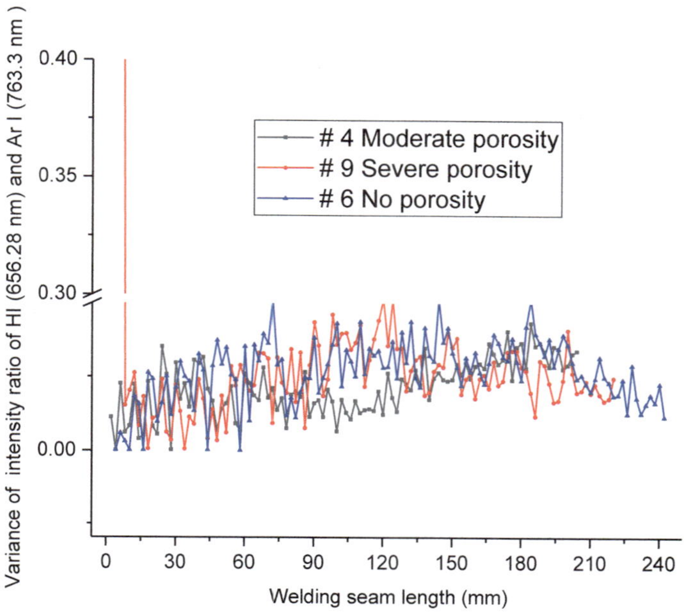

Fig. 2.23 Monitoring curves using the variance of spectrum intensity ratio of H I (656.28 nm) and Ar I (763.3 nm) for different levels of porosity seams

Table 2.5 Variables for constructing feature parameters for porosity detection

Variable	Value			Feature
Principal components	$Y_j = W^T x_i$ (j = 1)	$Y_j = W^T x_i$ (j = 2)	$Y_j = W^T x_i$ (j = 3)	Different coefficient of H spectrum
Sampling number	10	30	60	Different length of seam
Statistical feature	Variance	Kurtosis	Skewness	Different feature of curve

proposed for porosity monitoring and is described as

$$\mathrm{Var}(Y) = \frac{1}{n}\sum_{i=1}^{n}\left(Y_j - \mu\right)^2, \tag{2.3}$$

where $Y_j = W^T x_i$, $(j : 1, \ldots P; i : 1, \ldots 1322)$, $\mu = \frac{1}{n}\sum_{i=1}^{n} Y_j$.

In Eq. (2.3), the columns of W are the eigenvalues, ei, e.g., coefficient value, in descending order. Y is the new PC of the line spectrum, p is the number of PCs, and i is the dimension of the spectrum signal. Three kinds of welding seams with different levels of porosity were applied to validate the proposed method.

Figure 2.24 illustrates the feature curves corresponding to three distinct welding seams exhibiting varying degrees of porosity. Notably, in Fig. 2.24a and b, the features extracted from both PC2 and PC3 demonstrate an ability to delineate the severe

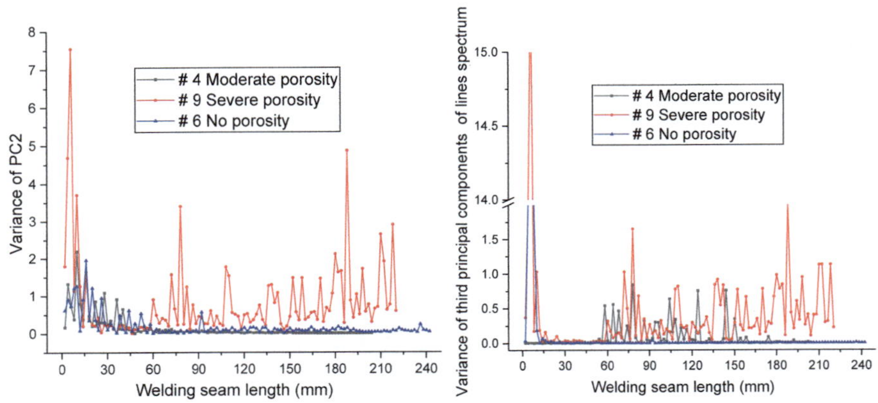

(a) Second PCs (b) Third PCs

Fig. 2.24 Monitoring curves of VarPCLS for three different porosity seams

porosity observed in Seam #9 from the moderate and non-porous seams. However, a closer examination reveals that the variance of PC3, as depicted in Fig. 2.24b, exhibits superior performance and heightened sensitivity in distinguishing between Seam #6 with no porosity and Seam #4 with moderate porosity. Interestingly, the feature value associated with Seam #6 in Fig. 2.24a appears higher than that of Seam #4, a discrepancy that contradicts the porosity volume detected via X-rays in Fig. 2.15 This inconsistency underscores the complexity inherent in porosity detection and highlights the need for more nuanced feature parameters. The proposed feature parameter, VarPCLS, predominantly reflects the comprehensive changes in metal elements (such as Al and Mg) as well as hydrogen (H). Consequently, it becomes imperative for VarPCLS to mitigate the influence of other chemical elements besides H, which are considered less impactful on porosity occurrence. By prioritizing H as the primary factor, VarPCLS aims to enhance its efficacy in accurately capturing variations in porosity levels during welding processes.

The presence of hydrogen (H) in aluminum alloy welding is closely associated with the formation of hydrogen porosity. This correlation stems from the abundance of water molecules within the refractory oxidation layer, leading to significant H element emissions in arc welding spectra. As the dynamics of the H spectrum undergo notable changes, they are presumed to be closely intertwined with the generation of inner porosity. In this context, the calculation of PC coefficients via PCA serves as a valuable tool for evaluating the dynamic characteristics of selected line spectra, including H I, Mg I, among others. To elucidate the relationship between PC coefficients and inner porosity, the coefficient values of the H atom spectrum (656.28 nm) were extracted and analyzed across seams exhibiting varying levels of porosity. Additionally, a comparison was drawn with the Mg I (383.83 nm) line spectrum, whereby the coefficient vector was computed for the entire seam. To ensure an adequate amount of information for porosity occurrence and a sufficient degree of H porosity detection, the number of PCs was set to 20. The resulting coefficient curves

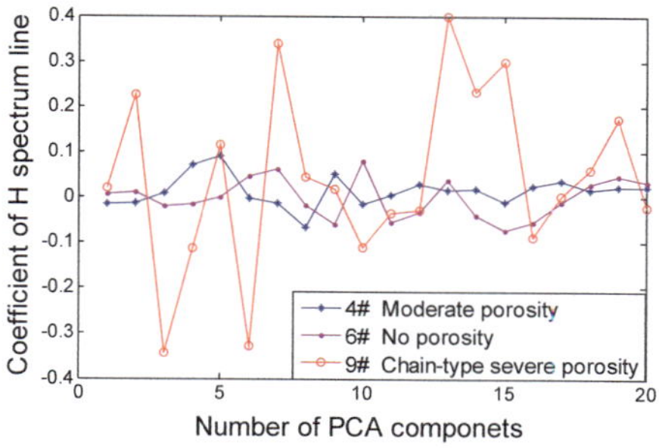

(a) H I spectrum line (656.28 nm)

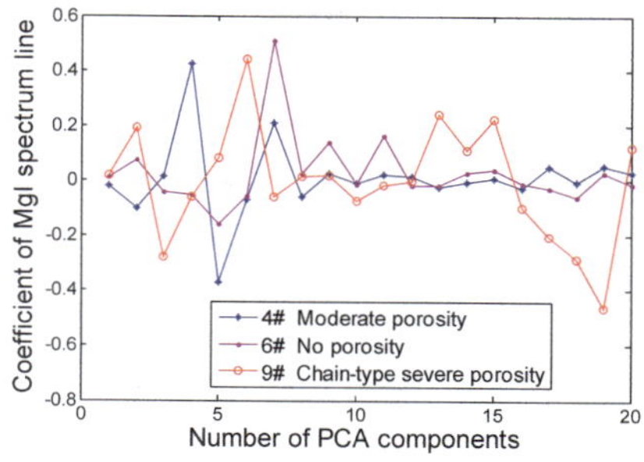

(b) Mg I spectrum line (383.83 nm)

Fig. 2.25 Coefficient curves for different levels of porosity for welding seams

for the two elements are depicted in Fig. 2.25, providing insights into their respective dynamics and their potential correlation with inner porosity levels. Through this analysis, a deeper understanding of the interplay between spectral features and porosity formation in aluminum alloy welding is anticipated.

Observing the coefficient curves depicted in Fig. 2.25a for distinct levels of porosity, it becomes evident that the coefficient curve associated with the H I spectrum of Seam #9 exhibits more pronounced fluctuations compared to seams #4 and #6, a trend consistent with the findings of X-ray testing. Conversely, the curves corresponding to seams #4 and #6, as well as those of the Mg I line spectrum illustrated in Fig. 2.25b, demonstrate notable similarity. To further assess the quantitative correlation in detecting various levels of porosity, statistical values were computed for the two sets of curves. The results are presented in Table 2.6, wherein “ei” represents the eigenvector of the PCA, and “p” denotes the number of pixels for each spectrum wavelength. These statistical metrics offer valuable insights into the efficacy of H I and Mg I spectra in discerning different porosity levels, facilitating a comprehensive evaluation of their utility in porosity detection during aluminum alloy welding processes.

To facilitate a comprehensive comparison of the performance of these two chemical elements, Fig. 2.26 presents a bar graph illustrating the statistical results. From Fig. 2.26a, it becomes apparent that all the statistical values associated with Seam #9 were markedly higher compared to the other seams, attributable to its chain-type severe porosity. This suggests that the coefficient of the H I spectrum effectively detects severe porosity, as evidenced by the elevated statistical values. In contrast, due to the relatively smaller increase in porosity within Seam #4 compared to Seam #6, only the maximum value displayed a noticeable increment. Notably, the range and standard deviation exhibited superior performance in differentiating between various levels of porosity, as depicted in Fig. 2.26b. However, it is worth noting that the Mg I spectrum displayed less stability in detecting severe porosity, characterized by lower maximum and mean values, whereas the H I spectrum exhibited better sensitivity and correlation to inner porosity. These findings underscore the efficacy of the H I spectrum in discerning porosity levels, particularly in severe cases, while highlighting

Table 2.6 Statistics of coefficient vector of H I and Mg I spectrum for seams with different levels of porosity

Statistics of coefficient vector	Coefficient of H I (p = 530)			Coefficient of Mg I (p = 64)		
	#6	#4	#9	#6	#4	#9
Max = $\text{maximum}\left(e_i^P\right), (i:1,\ldots 20)(p:pixle)$	0.079	0.09	0.397	0.51	0.424	0.442
$Mean = \frac{1}{n}\sum_{i=1}^{n} e_i^P, (i:1,\ldots 20)$	– 0.002	0.014	0.043	0.028	0.007	– 0.004
$Std = \sqrt{\frac{1}{n}\sum_{i=1}^{n}\left(e_i^P - Mean\right)^2}, (i:1,\ldots 20)$	0.044	0.034	0.198	0.133	0.143	0.207
$Range = \text{Max}(e_i^P) - \text{Min}(e_i^P), (i:1,\ldots 20)$	0.153	0.157	0.742	0.668	0.796	0.906

the limitations associated with utilizing the Mg I spectrum for this purpose. Such insights are instrumental in refining porosity detection methodologies and enhancing the efficacy of monitoring processes in aluminum alloy welding applications.

In a real manufacturing scenario, the length of a defective seam can vary unpredictably, necessitating adjustments to the input length of the principal component analysis (PCA) to encompass a wider seam area and effectively detect all potential porosity instances. Accordingly, the sampling number of a spectrum.txt file was set to 60, corresponding to a seam length of approximately 4 mm, thereby fulfilling the practical requirements of porosity detection. Subsequently, the coefficient of the H I line spectrum for the first principal component was extracted, and the monitoring curves of the welding seams were depicted in Fig. 2.27a. To ascertain the capability of the proposed feature in monitoring different porosity levels, statistical features were computed from the feature curves in Fig. 2.27a using the algorithms outlined in

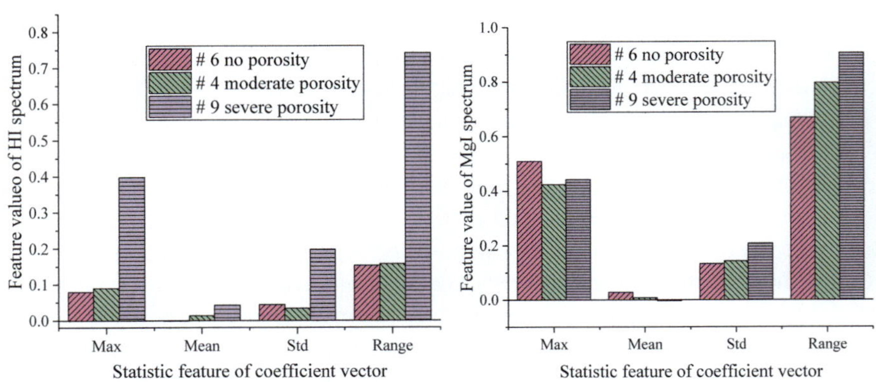

(a) H I spectral line (656.28 nm) (b) Mg I spectral line (383.83 nm)

Fig. 2.26 Statistics for detecting different levels of porosity for welding seams in Fig. 2.15

Table 2.6. The region of interest for statistical analysis is delineated by a dark rectangle, encompassing varying levels of porosity generated within seams #6, #4, and #9. Their corresponding X-ray film and statistical results are presented in Fig. 2.27b and c, respectively. It is evident from these figures that all statistical features increased as the severity of porosity worsened, except for the median porosity, suggesting that the coefficient of the H I line spectrum from the first principal component effectively reflects the level of porosity. Furthermore, the absolute value of the coefficient was determined for each data point in Fig. 2.27a to enhance the effectiveness of the proposed feature. Figure 2.27d illustrates the feature curves based on the absolute value, which clearly differentiate between the three types of porosity, particularly for Seam #4 and Seam #6. Therefore, it can be concluded that the absolute value of the coefficient of the H I line spectrum from the first principal component serves as a highly effective monitoring parameter for detecting inner porosity during dynamic welding processes.

The T-SNE (t-distributed stochastic neighbor embedding) tool, pioneered by Hinton in 2008, stands as a formidable asset in data reduction and visualization, particularly for high-dimensional datasets. Its prominence rivals that of deep learning methodologies in research circles focused on manifold learning. D. Wu's work, is a notable example wherein t-SNE, when integrated with a deep belief network, facilitated the identification of seam penetration status during polarity plasma arc welding processes, achieving both feature dimension reduction and visualization.

T-SNE represents an advancement over the original Stochastic Neighbor Embedding (SNE) technique, which was developed to address the challenges of nonlinear feature reduction. With its ability to effectively capture nonlinear relationships within data, T-SNE has emerged as a cornerstone in manifold learning methodologies, offering researchers powerful tools for exploring and understanding complex datasets. It defines the high-dimensional Euclidean distances between data points as conditional probabilities. The similarity of input sample pairs is the conditional probability, pjli, and is given by

$$p_{j|i} = \frac{\exp\left(-\|x_i - x_j\|^2 / 2\sigma_i^2\right)}{\sum_{k \neq i} \exp\left(-\|x_i - x_j\|^2 / 2\sigma_i^2\right)} \tag{2.4}$$

where i and j are the index variables and σi is the variance of the Gaussian distribution centered on data point xi. Likewise, in the low-dimensional feature space, we can compute a similar conditional probability, qjli. The Gaussian variance was set to. Hence, we define the probability distribution as

$$q_{j|i} = \frac{\exp\left(-\|y_i - y_j\|^2\right)}{\sum_{k \neq i} \exp\left(-\|y_i - y_k\|^2\right)}. \tag{2.5}$$

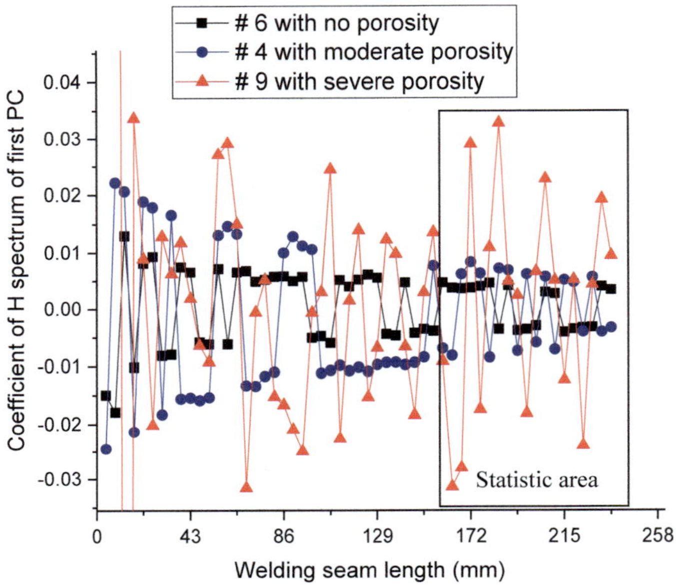

(a) Coefficient of H I spectrum from first PC.

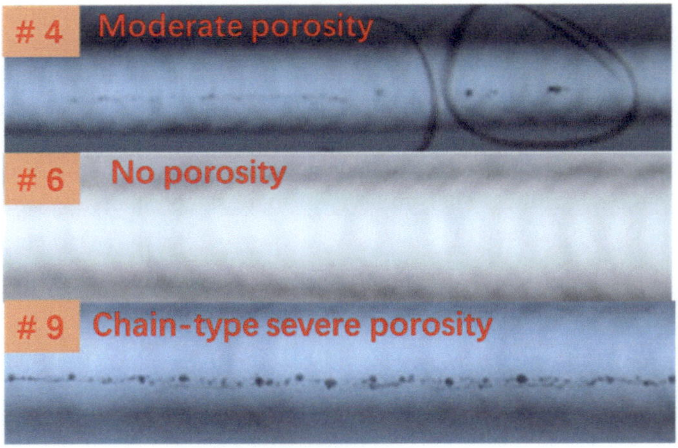

(b) Enlarged seam for statistical area

Fig. 2.27 Proposed spectrum features for detecting different levels of porosity based on PC coefficient

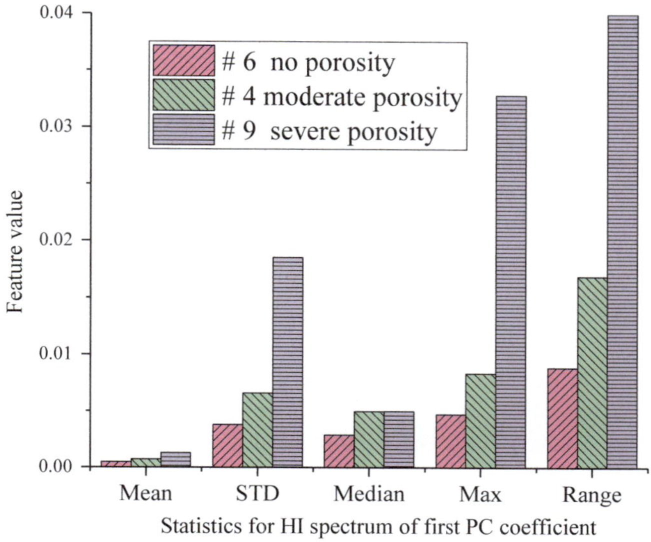

(c) Statistical results.

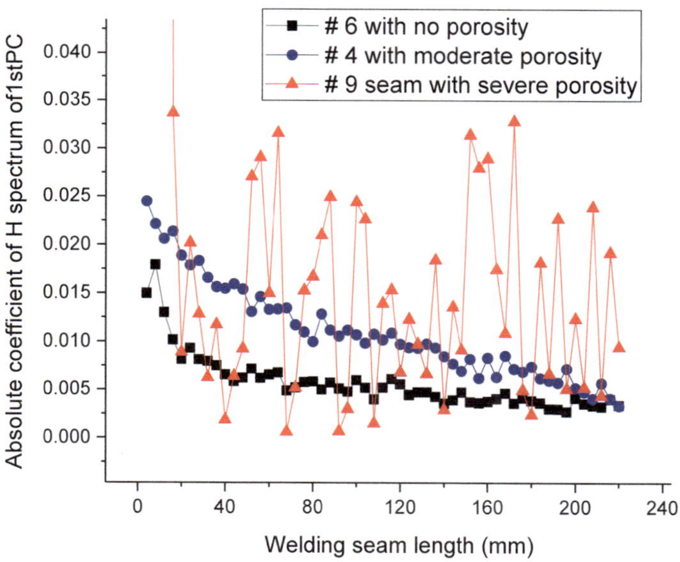

(d) Absolute coefficient of H I line spectrum of first PC.

Fig. 2.27 (continued)

SNE tries to match these two distributions. If the map points indicate that the sample pairs (yi, yj) from the low-dimensional space are similar to the high-dimensional sample pairs (xi, xj), matrices pj|i and qj|i will be equal. Motivated by this observation, in SNE, Kullback–Leibler divergence is used to measure the faithfulness with a gradient descent method to minimize the mismatch between pj|i and qj|I. The cost function C is defined as

$$C = \sum_i KL(P_i \| Q_i) = \sum_i \sum_j p_{j|i} \log \frac{p_{j|i}}{q_{j|i}}, \tag{2.6}$$

where Pi and Qi are the conditional probability distribution over all other data points given data point xi and all other map points given map point yi, respectively. To solve the crowding problem of SNE, t-SNE applied a student t-distribution with one degree of freedom instead of a Gaussian distribution. Using the t-distribution, the joint probabilities qj|i are defined as

$$q_{j|i} = \frac{\left(1 + \|y_i - y_j\|^2\right)^{-1}}{\sum_{k \neq i} \left(1 + \|y_i - y_j\|^2\right)^{-1}}. \tag{2.7}$$

The gradient of the Kullback–Leibler divergence is

$$\frac{\delta C}{\delta y_i} = 4 \sum_j \left(p_{ij} - q_{ij}\right)\left(y_i - y_j\right)\left(1 + \|y_i - y_j\|^2\right)^{-1}. \tag{2.8}$$

In this section, PCA and tSNE were integrated to reduce the PCA dimension and improve the ability to distinguish different classes of porosity seams. The flowchart of PCA-tSNE is displayed in Fig. 2.28.

In Fig. 2.29a, the two-dimensional feature space generated using the PCA coefficient of the H spectrum reveals a limited separation between different classes of porosity. However, upon applying tSNE to the original PCA feature data, the visualization results depicted in Fig. 2.29b showcase distinct separation among the three types of porosity seams. This underscores the potent capability of the proposed PCA-tSNE method in both feature reduction and visualization. The efficacy of this approach extends beyond porosity detection and can be leveraged across various monitoring signals to enhance scientific understanding of complex manufacturing processes and optimize the utilization of monitored data. By integrating PCA with tSNE, researchers gain access to a robust methodology for uncovering hidden patterns and relationships within multidimensional datasets, thereby enabling more informed decision-making and process optimization in industrial settings.

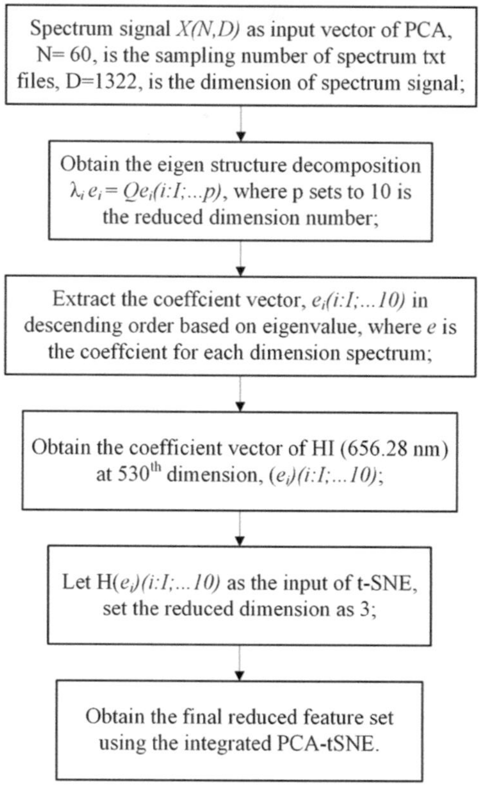

Fig. 2.28 Flowchart of proposed feature reduction method of PCA-tSNE

2.3.4 Correlation Between Microscopic Porosity and Spectrum Feature

To enhance the visibility of inner porosity, a predetermined depth of gap was intentionally created along the direction of a seam, manifested in two distinct forms on the left side of the specimen and delineated by circles. Subsequently, tensile tests were conducted utilizing a CSS-88100 universal test machine, employing a drawing speed of 1 mm/min, followed by observation of the fracture section using a VEGAIIXMU scanning electron microscope. To further analyze the composition within, outside, and surrounding the porosity, an energy dispersive spectroscope (OXFORD- 7718) was employed. Table 2.7 provides a comprehensive overview of all the samples subjected to microscopic porosity testing, outlining the experimental setup and parameters employed in this study.

The non-destructive testing (NDT) results obtained from welded Seam #4 in Fig. 2.30a reveal a higher incidence of inner porosity, particularly notable when the surface oxidation film was left intact on the second half of the seam. Additionally, scattered porosity sites were detected, accompanied by narrower seams exhibiting under-penetration at the onset of the seam. This occurrence suggests that insufficient

Fig. 2.29 Feature reduction and visualization using PCA-tSNE

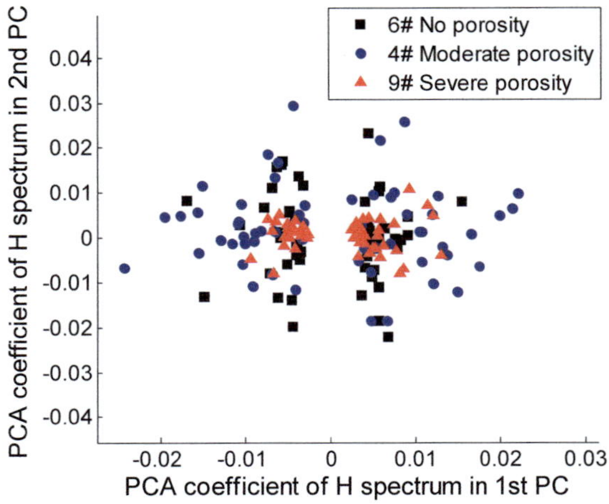

(a) Original PCA features.

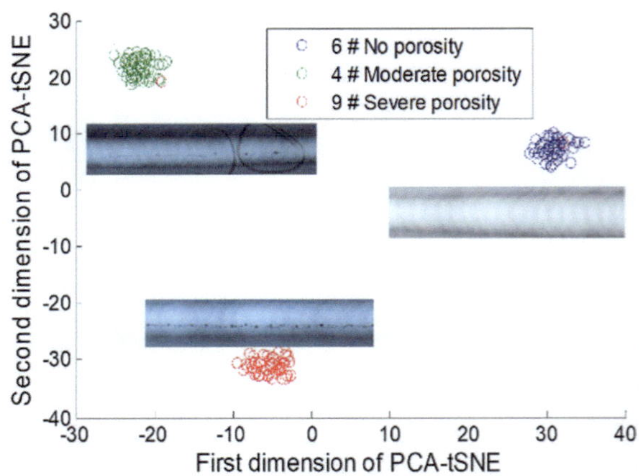

(b) Feature distribution by PCA-tSNE.

Table 2.7 Samples for microscopic porosity testing

Weld seam	Tested samples	Seam condition	Testing
# 4	4-1, 4-2 4-3, 4-4	Oxidation film removed, oxidation film non-removed	Tensile test Fracture analysis EDS testing
# 6	6-1, 6-2 6-3, 6-4		
# 9	9-1, 9-2		

penetration may have hindered the conditions necessary for porosity to escape or prevented its formation and growth. A comparison between Fig. 2.30b and c highlights fewer and smaller porous areas observed on fracture Section F-F in contrast to Section C-C. Notably, upon closer examination of the enlarged images, several scattered porous sites were identified on the bottom of the seam. The mean diameter size of porosity, as depicted in Fig. 2.18b, measured at 54 μm, was smaller compared to the mean size of porosity observed in fracture Section C-C, which measured at 93.21 μm, as evident in the enlarged image of D. Furthermore, Fig. 2.30c illustrates the specimen fracture occurring along the chain-type porosity, effectively exposing various sizes of porosity. These observations underscore the critical role of surface preparation and penetration depth in mitigating porosity formation and growth during the welding process. Additionally, they emphasize the importance of comprehensive inspection techniques, such as NDT and fractography, in assessing weld quality and identifying potential defects.

To establish a more robust correlation between the arc spectrum feature and the micro-characteristics of inner porosity, Fig. 2.30d presents the feature curve depicting the intensity ratio between H I (656.28 nm) and Ar I (763.3 nm) denoted by the black square-dot line. In contrast, the red dot line represents the feature curve of the third principal component (PC), primarily derived from the metal line spectrum of welding Seam #4, as illustrated in Fig. 2.30a. An evident pattern emerges wherein the fewer and smaller porous sites observed in fracture Section F-F align with lower feature values of H/Ar in Fig. 2.30d. Conversely, seam Section C-C, characterized by more porosity, exhibits a higher feature value, indicative of increased porosity generation within the seam featuring surface oxidation film. However, it's noteworthy that the behavior of the third PC of the line spectrum contradicts this microscopic porosity rule. Notably, its lower feature value corresponds to the fracture Section of C-C, despite exhibiting more porosity. This discrepancy suggests that the generation rule of porosity may be significantly influenced by the behavior of the metal component, such as Al and Mg, predominantly reflected in the third PC constructed mainly by the metal line spectrum. Furthermore, the oxidation film's presence exerts a cooling effect on the welding arc by absorbing the arc's heat for melting and ionization, thereby influencing the welding process dynamics.

Based on the energy dispersive spectrometer test results outlined in Table 2.8, it was observed that magnesium (Mg) plays a significant role in both the generation and expansion of porosity during the dynamic welding pool evolution in gas tungsten arc welding (GTAW) of aluminum (Al) alloy. Consequently, two distinct types of porosity emerge within the welding seam of Al alloy during GTAW: Mg vapor porosity and hydrogen (H) porosity. These are primarily instigated by the presence of non-fusible materials, either from Mg oxidation or hydrogen, along with the rapid and uneven transition of the welding pool from liquid to solid state. Table 2.6 provides the mean Mg content measured within and outside both types of porosity, with the mean values calculated from four respective points. Notably, for H porosity, the average Mg content within the inner wall of the porosity was lower compared to that outside, irrespective of the oxidation film treatment. This disparity suggests that H porosity may stem from the accumulation, reaction, and subsequent formation of hydrogen gas

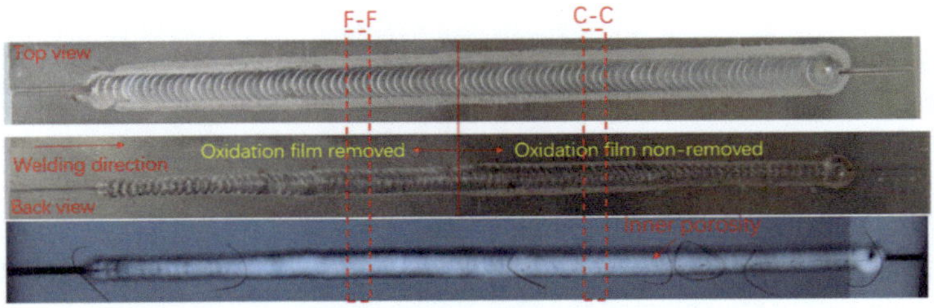

(a) Welding Seam #4 and its X-ray result.

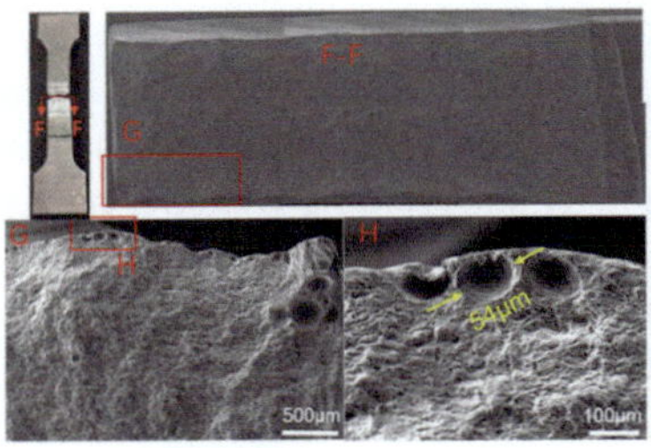

(b) Oxidation film removed.

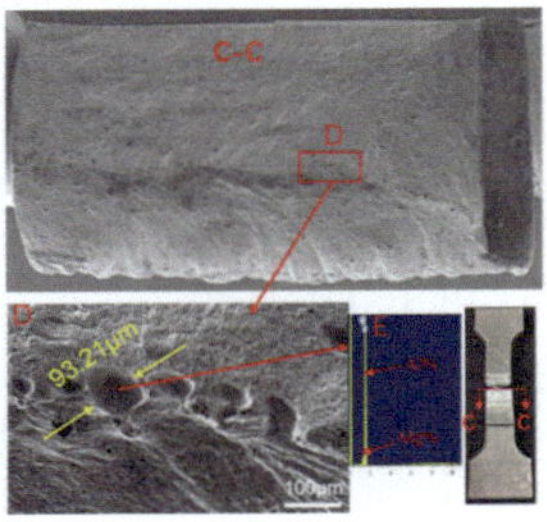

(c) Oxidation film not removed.

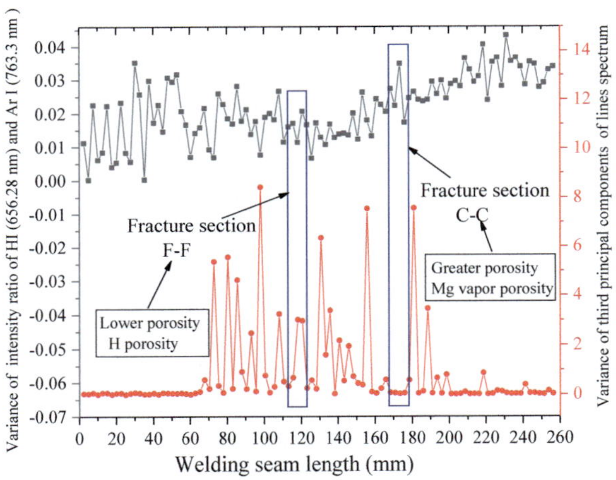

(d) Feature curves of H/Ar and third PC of metal line spectrum for welding Seam #4 in (a).

Fig. 2.30 Tensile fracture section correlation to spectrum feature

(H2). Moreover, the abundance of Mg outside the porosity was notably higher in fracture Section C-C when the oxidation film remained intact, indicating that magnesium oxide from the refractory likely contributes to the elevated Mg levels surrounding the porosity. In contrast, for Mg vapor porosity, the mean Mg content within the porosity was 5.14%, exceeding that found outside. This observation suggests that the accumulation, vaporization, and subsequent entrapment of magnesium vapor within the

Table 2.8 Mean Mg content of two types of porosity tested in different areas (%) using EDS

Type of porosity	Tested seam	Inner wall of porosity	Outside of porosity
H porosity	Fracture section C-C (oxidation film not removed)	2.1	4.8
	Fracture section F-F (oxidation film removed)	1.65	3.74
Mg vapor porosity	Fracture section C-C (oxidation film not removed)	5.14	2.5

porosity contribute to its elevated Mg content. The correlation between microscopic porosity analysis and spectrum features is evident. H porosity exhibits a positive correlation with the intensity ratio of H I/Ar I, with larger porosity volumes corresponding to higher H/Ar ratios. Conversely, a negative correlation exists between porosity, whether H or Mg vapor, and the metal spectrum, where a decrease in metal spectrum value correlates with larger porosity volumes. These findings underscore the intricate relationship between alloy composition, welding process dynamics, and the resultant porosity morphology, highlighting the importance of understanding and controlling these factors in optimizing weld quality and integrity.

2.3.5 *Inner Porosity Detection Results*

When comparing the traditional approach of utilizing the spectrum intensity ratio of H I/Ar I, as depicted in Figs. 2.22 and 2.23, with the proposed detection features—namely, the statistics of PCA-PCs illustrated in Fig. 2.24 and the coefficient of the H line spectrum shown in Fig. 2.27a notable improvement in results is observed. This enhancement suggests that PCA effectively extracts intrinsic features from the spectrum data that correlate with changes in welding conditions. Specifically, the dynamic changes of the H I line spectrum are quantitatively characterized, providing valuable insights into porosity dynamics. Importantly, the utilization of line spectrum features is independent of the absolute intensity of spectrum emission, offering a non-dimensional parameter that enhances robustness and stability for future industrial applications.

The occurrence of inner porosity presents a significant challenge for real-time detection due to its high uncertainty and the inherent lag between porosity generation and real-time spectrum emission. Detecting and accurately locating inner porosity during dynamic welding processes remain formidable tasks. In this study, a detection length of 4 mm was deemed sufficient for practical demands, yet there is a pressing need for future work to enhance detection accuracy for inner porosity.

Our findings regarding porosity distribution suggest that scraping the bottom or top layer of a seam with suspected porosity may passively reduce seam inner porosity. Moving forward, to advance the on-line defect detection technology for widespread industrial applications, it is imperative to align the relevant standards of welding quality and defect detection with current research endeavors. This concerted effort will facilitate the seamless integration of cutting-edge research findings into industrial practices, thereby enhancing overall welding quality and efficiency.

2.3.6 Conclusion

This investigation delved into the realm of on-line inner porosity detection for Al-Mg alloy through robotic Gas Tungsten Arc Welding (GTAW), utilizing post-testing methodologies alongside arc optical spectrum signals. From our comprehensive analysis, the following key conclusions emerge:

(1) Effect of Surface Oxidation Film: The presence of the surface oxidation film significantly influences the detection of inner porosity in Al alloy during pulsed GTAW. Notably, X-ray testing revealed a greater presence of inner porosity when the oxidation film was retained. Employing Principal Component Analysis (PCA) on both background and line spectra unveiled a selection of metal, Ar, and H spectra closely intertwined with the dynamic welding process.
(2) Development of Feature Parameters: Feature parameters derived from the newly generated Principal Components (PCs) and PCA coefficients proved instrumental in porosity detection during the Al alloy welding process. Among these, the absolute coefficient of the H I line spectrum from the first PC demonstrated superior performance in discerning seams with varying degrees of porosity, ranging from none to moderate to chain-type severe. Furthermore, the application of PCA-tSNE facilitated effective dimension reduction and visualization of diverse porosity levels.
(3) Correlation Between Porosity and Spectrum Features: Intriguingly, a correlation between microscopic porosity analysis and spectrum features emerged. Two distinct types of inner porosity were identified for Al alloy in GTAW: Mg vapor porosity and H porosity. H porosity exhibited a positive correlation with the intensity ratio of H I/Ar I, with higher porosity volumes corresponding to elevated H/Ar ratios. Conversely, a negative correlation was observed between porosity—be it H or Mg vapor—and the metal spectrum, where larger porosity volumes correlated with lower metal spectrum values.

These findings not only enhance our understanding of porosity dynamics during Al alloy welding but also pave the way for the development of robust and efficient on-line detection methods crucial for ensuring welding quality and integrity in industrial settings.

2.4 Inner Porosity Monitoring Based on Advanced Spectral Features

2.4.1 Experiment Design

The experimental system comprises a robotic welding system, a multi-information synchronous acquisition system, and a process control system. Figure 2.31a illustrates the main components of the welding system. Figure 2.31b depicts a welding scene captured by the multi-information acquisition system. Additionally, control over welding current, wire feed speed, and welding speed is achieved using LabVIEW software and data acquisition cards, enabling synchronous triggering of multiple signal sensors. Synchronous acquisition and storage of current, voltage, sound, images, and arc spectra are performed. Arc spectra are collected in real-time using a fiber optic spectrometer (AvaSpec-1350F-USB2) fixed at the end of the robot arm, covering a range from 350 to 1100 nm, with a CCD pixel count of 1350, an integration time set at 1.5 ms, and a sampling rate of approximately 70 Hz. The positioning of the probe is highly sensitive, necessitating pre-experimental adjustment by moving the probe and real-time spectral acquisition to analyze and select an optimal probe position. Ultimately, with a certain level of arc spectrum intensity ensured and minimal obstruction by the tungsten electrode, the probe is positioned directly at the center of the arc, with a 10% neutral density filter added to maintain the spectrum intensity within around 2/3 of its full scale.

Welding of 5A06 Aluminum Alloy with a 4 mm Thick Y-groove Butt Joint, utilizing specific welding parameters as outlined in Table 2.1. Aluminum alloys are prone to forming a dense oxide film on their surfaces, which readily combines with water molecules in the air, serving as the primary hydrogen source for porosity formation. In this experiment, the oxide film is considered the main interfering factor, simulating the creation of porosity defects. Table 2.10 presents the experimental design for porosity defects. Post-welding, the weld seams are analyzed and characterized. Samples are polished, corroded, and examined using a Nikon Eclipse MA200 optical microscope for both transverse and longitudinal cross-sections. To expose more porosity upon tensile fracture, notches are pre-installed at the edges of the weld seam. Tensile tests are conducted at a speed of 1 mm/min using a CSS-88100 tensile machine. The fracture surfaces are observed for porosity morphology, distribution, and size using a VEGAIIXMU scanning electron microscope, while the internal and external compositions of the pores are determined using an OXFORD-7718 energy dispersive X-ray spectrometer (Table 2.9).

Figure 2.32 presents the weld seam with porosity intentionally created by leaving the oxide film uncleared, along with the results of the non-destructive testing. It is evident that within the uncleared oxide film region in the latter part of the weld seam, chain-like porosities are observed. Additionally, some scattered porosities are detected at the narrower fusion zone of the weld seam during the initial welding stages. It is widely reported that porosity formation is a consequence of the interplay between porosity buoyancy and solidification of the liquid weld pool, with

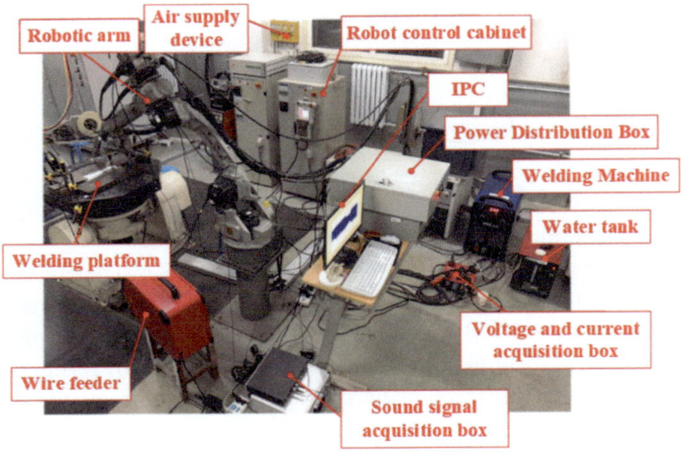

(a) The Robotic Welding Experimental System.

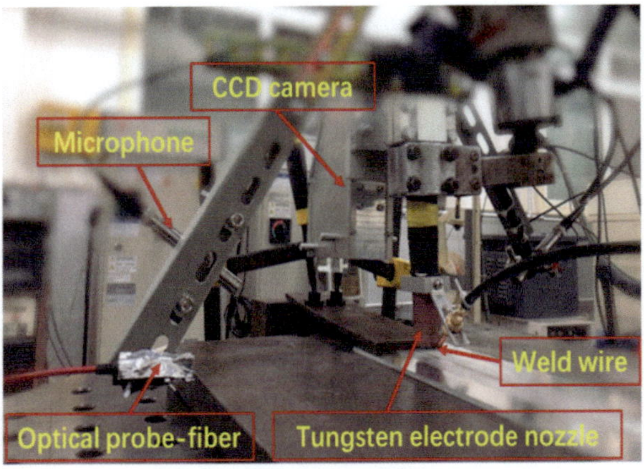

(b) Multi-source Information Acquisition Physical Diagram.

Fig. 2.31 Multi-source sensor information acquisition for welding

Table 2.9 Welding parameters

Parameter	Value	Parameter	Value
Pulse frequency (Hz)	1	Welding speed (cm/min)	16
Peak current (A)	240	Wire feeding speed (mm/s)	12
Base current (A)	50	Tungsten electrode diameter (mm)	3.2
Duty cycle (%)	50	Welding wire model/diameter (mm)	SGAlMg6/1.6
Argon gas flow rate (L/min)	15	Sheet size (mm)	300 × 50 × 4

Table 2.10 Welding experiment design and analysis testing

No.	Experiment design	Post weld testing
1	Completely clean the oxide film and connect the Y-shaped groove	X-ray non-destructive testing, cross-section and longitudinal observation Hardness testing of the weld cross-section, pre-notched tensile testing, SEM observation of the fracture surface for porosity morphology, and EDS composition determination
2	Clean the oxide film in the first half and do not clean it in the second half. Y-shaped groove docking	
3	The entire section is not cleaned of the oxide film, and the Y-shaped groove is docked	
4	Clean the oxide film in the first half and do not clean it in the second half. V-shaped groove flat plate overlay welding	

weld penetration and cooling rate being the primary influencing factors. To investigate the distribution patterns and mechanisms of porosity formation, observations were conducted on different cross-sections of the weld seam, and tensile specimens were prepared for porosity observation and compositional analysis, as outlined in Table 2.11.

Figure 2.33 contrasts the cross-sectional images of weld seams with different surface treatments. It can be observed that porosities are primarily distributed at the top and bottom of the weld seam, with some scattered porosities near the fusion line. In comparison, Fig. 2.33a exhibits a higher density of porosities, potentially

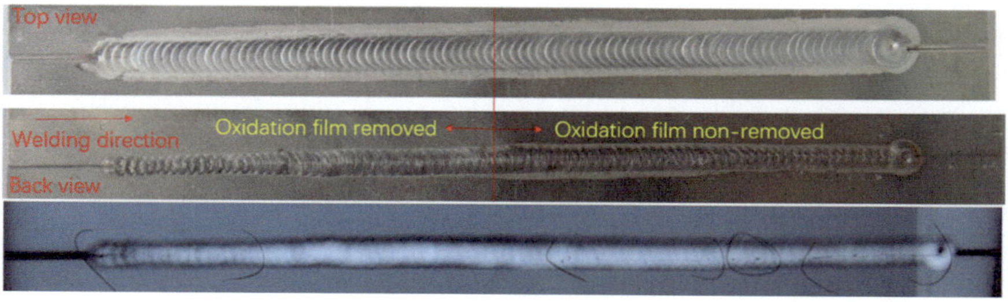

Fig. 2.32 Weld seam no. 2 and non-destructive testing results

Table 2.11 Test specimen preparation

No.	Surface treatment	Test	Cross section observation
1	Remove surface oxide film (OFR)	Tensile testing (TS) Hardness testing (HS)	Microscopic cross-section (MT)1
2			
3	Preserve surface oxide film (OFNR)		Microscopic cross-section (MT)2
4			Longitudinal section (LS)3

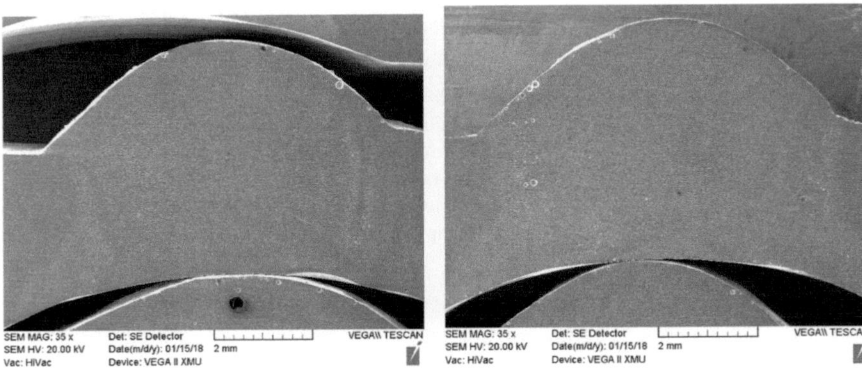

(a) MT4-1OFR (b) MT4-2 OFNR

Fig. 2.33 Cross section of no 4 group weld seam

attributed to the faster cooling rate near the fusion line. The presence of porosities at the bottom may be linked to inadequate shielding gas protection. In practical welding production, increasing the cooling rate can be employed to suppress porosity growth. Furthermore, surface grinding of the weld seam can be utilized to eliminate potential porosity defects, thereby enhancing the quality of the weld seam.

For the same weld seam, a longitudinal section A-A was extracted, polished, and observed, as depicted in Fig. 2.4. Similar to the weld seam cross-section in Fig. 2.33, porosities are primarily distributed at the bottom of the weld seam. In the magnified view labeled as B, spherical internal smooth porosities can be observed, which are attributed to hydrogen gas pores formed due to differences in hydrogen solubility between the liquid weld pool and solidified weld pool, consistent with findings reported in other literature. Additionally, adjacent to the porosities, unfused defects can be observed, possibly caused by refractory oxide films. From the perspective of porosity nucleation, refractory oxide films can provide certain conditions for the nucleation and growth of porosities.

Comparing the distribution and composition of porosities in weld seams after tensile fracture with different surface treatments. Notches were pre-cut along the sides of the tensile specimens using a wire cutting machine to expose as many porosities as possible upon fracture, as shown in Fig. 2.5. The fracture occurred along the centerline C-C of the weld seam, where the pre-cut notches are clearly visible, and the specimen fractured along the chain-like porosities. Magnification of area D, where chain-like porosities are present, reveals spherical smooth-walled porosities. EDS analysis was conducted to determine the composition of the interior and exterior of the porosities, with results summarized in Table 2.4.

The interior and exterior of the porosities were each sampled four times, and the average values were calculated. From Table 2.12, it can be observed that the average Mg content on the interior walls is 5.14%, approximately twice that of the exterior, indicating a certain degree of Mg enrichment within the porosities. Calculations for the larger porosities yield an average diameter of approximately 93.21 μm, while the smaller ones measure around 63.12 μm.

Observations and compositional measurements of porosity were conducted on the fracture surface of weld seams with cleaned oxide films, as depicted in Fig. 2.36. Compared to Fig. 2.35, there are significantly fewer porosities observed, with only scattered porosities present at the bottom of the weld seam, consistent with the observations from Figs. 2.33 and 2.34. Magnified views of the porosities at the bottom of the weld seam are shown in G and H. The average diameter of the porosities is calculated to be approximately 54 μm, smaller than those observed in Fig. 2.35. This demonstrates that welding without cleaning the oxide film can result in denser and larger porosities. Thus, complete removal of the oxide film is crucial for suppressing internal porosity defects in weld seams (Fig. 2.36).

Table 2.13 compares the Mg content in the interior walls and exterior edges of porosities under different surface treatments. Interestingly, the Mg content in the

Table 2.12 Composition analysis of porosities at TS4-4OFNR fracture surface (%)

Measurement points	C	O	Mg	Al	Cl	Fe	Dotting area
Point 1			5.85	78.37		15.8	Inner wall of stomata
Point 2			5.23	82.61		12.2	
Point 5		10.4	4.54	85.02			
Point 6			4.93	95.07			
Mean			5.14				
Point 7			2.04	97.96			Outer edge of stomata
Point 8			2.40	97.60			
Point 3	16.8	16.2	3.13	63.14	0.34	0.33	
Point 4			2.44	97.56			
Mean			2.5				

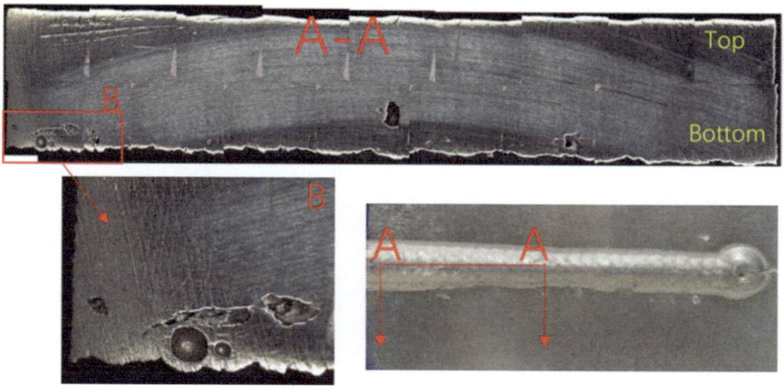

Fig. 2.34 Cross section of No4 group weld seam

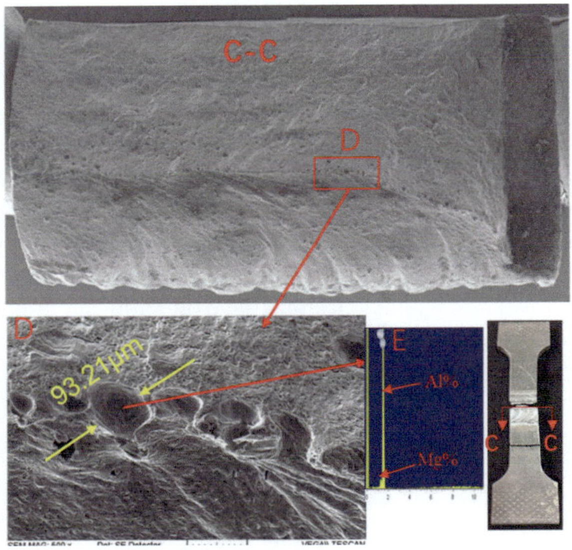

Fig. 2.35 Observation and measurement of tensile fracture porosity of uncleaned oxide film weld seam

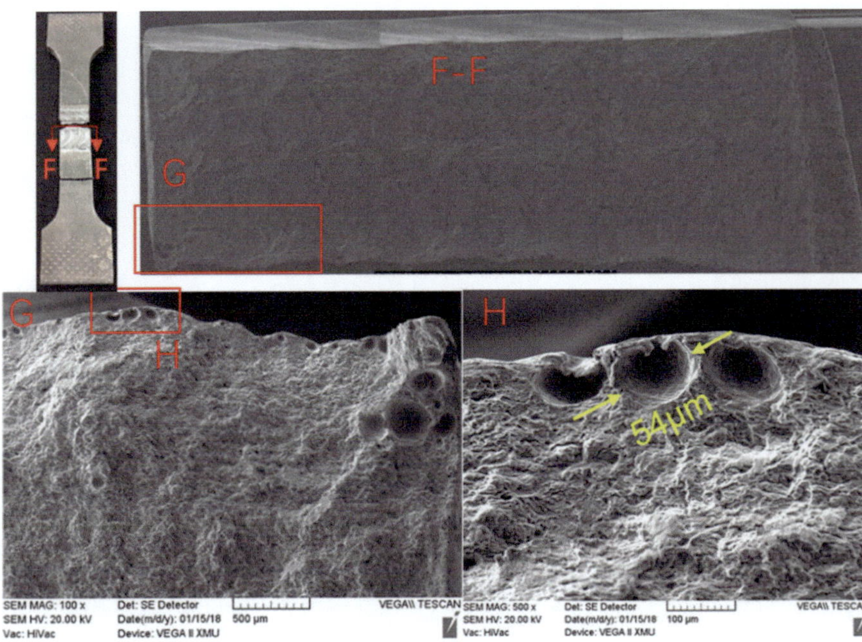

Fig. 2.36 Observation and measurement of porosities at the fracture surface of weld seam with cleaned oxide film

exterior edges is consistently higher than that in the interior walls, regardless of whether oxide film removal was performed. This result contradicts Table 2.12, indicating the significant role of Mg in the formation and growth of porosities. Due to the lower melting and boiling points of magnesium compared to aluminum, magnesium is prone to evaporating and forming magnesium vapor during the dynamic changes in the liquid weld pool. If magnesium vapor cannot escape the weld pool in time, it can lead to the formation of magnesium vapor porosities, explaining the enrichment of magnesium within the porosities as observed in Table 2.12. The compositional results obtained in Table 2.13 validate the presence of hydrogen gas porosities reported in the literature. Therefore, the GTAW welding process of aluminum alloys may produce two types of porosities: hydrogen gas porosities and magnesium vapor porosities. Both are caused by refractory hydrogen or oxide films, which are unable to escape during rapid cooling and significant liquid-to-solid phase transformations.

Figure 2.37 displays the morphology of exposed porosities at different magnification levels on the tensile fracture surface. Since the appearance of shrinkage cavities is very similar to that of porosities, areas suspected to be porosities were observed at a magnification of 800 times. Porosities were identified by their smooth inner walls, as shown in the right image. On the left side of Fig. 2.37, various-sized voids are observed, possibly indicating the aggregation and enlargement of porosities. The smallest void measures approximately 39.4 μm, while the larger aggregated porosity is about 125.9 μm.

Increasing the moisture content in the oxide film by soaking exacerbates porosity defects. Figure 2.38 provides SEM images of locally exposed porosities after tensile fracture, comparing the Mg content in the inner walls of labeled large and small porosities. It is found that the average Mg content in the inner walls of small porosities is approximately three times higher than that in large porosities. The elevated magnesium content in small porosities may originate from refractory oxide films, further confirming the significant involvement of magnesium in porosity formation during the solidification process of the weld pool.

Table 2.13 Mg content measurement (%) in porosities at the fracture surface

Tensile fracture specimen	Measurement points	Inner wall of stomata	Stomatal outer edge
TS4-3 OFNR (uncleaned oxide film)	1	2.00	4.02
	2	2.44	3.28
	3	1.87	7.11
	Mean	2.1	4.8
TS4-1OFR (clean the oxide film)	1	1.18	3.56
	2	1.87	3.94
	3	1.91	3.73
	Mean	1.65	3.74

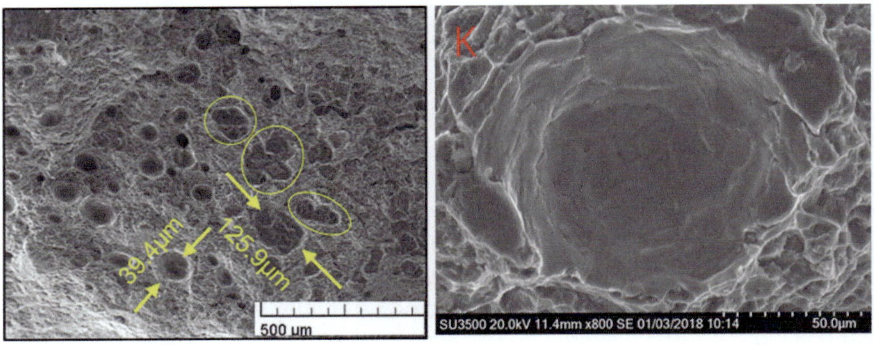

Fig. 2.37 Observation of porosities at the fracture surface of weld seam without oxide film cleaning

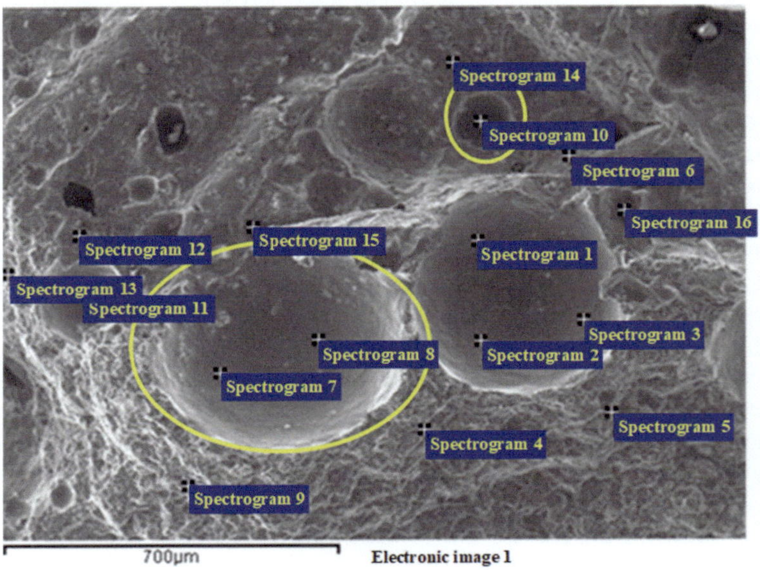

Fig. 2.38 Observation of porosities at the fracture surface of weld seam without oxide film cleaning after water soaking

2.4.2 Spectral Deep Mining Process

Electric arc spectroscopy plays a pivotal role as a sensing modality in the pulsed Gas Tungsten Arc Welding (GTAW) process for aluminum alloys. It provides essential insights into the welding process, offering unique cyclic variations that are influenced by various process parameters. The application of pulsed current introduces distinctive pulse characteristics into the process signals, as illustrated in Fig. 2.39. These pulse characteristics, including sound, voltage, and spectra, exhibit noticeable variations under different peak and base current conditions.

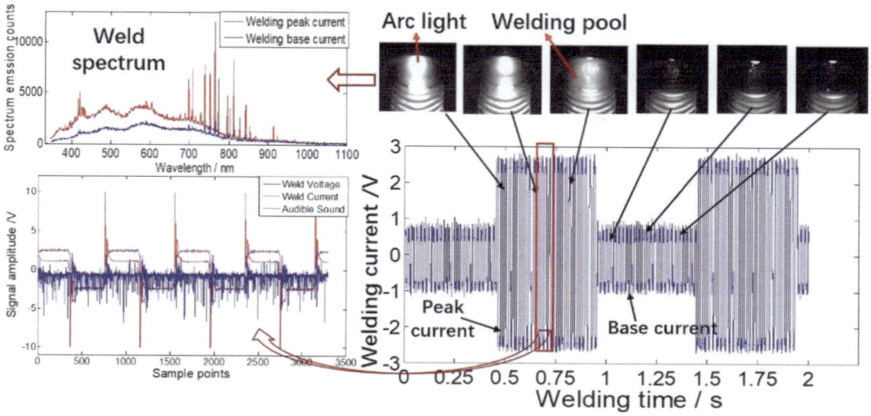

Fig. 2.39 Multi-source signals acquired during pulsed gas tungsten arc welding (GTAW) of aluminum alloys

This section focuses on an in-depth exploration of spectral signals within the GTAW process. It encompasses various analytical techniques such as feature dimensionality reduction and extraction. Furthermore, the correlation between spectral signals and the presence of porosity within weld seams is investigated. Additionally, efforts are made towards developing methods for online porosity detection and classification visualization. Figure 2.40 provides a visual representation of the proposed methodology, outlining the workflow involved in the comprehensive analysis of spectral signals for porosity detection and classification during the GTAW process.

2.4.3 Spectral Data Mining Based on PCA

Principal Component Analysis (PCA) is a common method for feature compression and dimensionality reduction, widely applied in fields such as machine learning, pattern recognition, and artificial intelligence. It achieves feature compression through linear transformation, aiming to represent the original feature information with as few dimensions as possible. The linear transformation process involves two key sets of parameters: eigenvectors (v) and eigenvalues (C). These parameters determine how high-dimensional features are assigned different coefficients and linearly combined to form various new principal components (PCs). Thus, eigenvectors can be regarded as quantitative assessment parameters for feature weights. Therefore, PCA serves not only as a dimensionality reduction tool but also facilitates in-depth feature mining of high-dimensional spectral information.

As depicted in Fig. 2.40, the initial step involves separating the line spectra from the background spectra, resulting in 1322-dimensional line spectral features. A complete weld seam spectral data is used as the input sample for PCA, with the

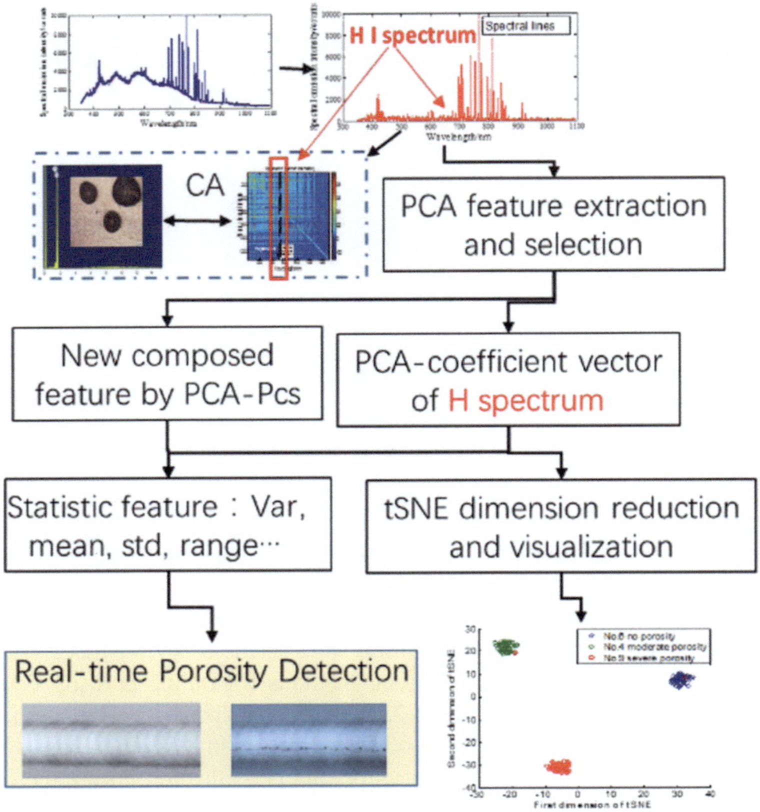

Fig. 2.40 Flowchart of spectral depth mining

sample length determined by the specific length of the weld seam. PCA processing is applied to the entire process line spectra signals of the unwashed weld seam subjected to water soaking, yielding the coefficient vectors of the first three principal components and their corresponding monitoring curves for the entire process, as illustrated in Fig. 2.41.

From Fig. 2.41a, it is evident that the amplitude of the first principal component (PC1) remains relatively stable under base current conditions, while it sharply decreases during peak current phases. Considering that the background spectra have been completely removed, coupled with Fig. 2.41b illustrating the coefficient plot of the first principal component, it is notable that the major component is attributed to the Argon spectrum. The absolute magnitudes of the PCA vector coefficients for the Argon spectrum are all greater than 0.1, indicating that the decreasing trend is primarily caused by the Argon spectral lines. Further analysis suggests that this decrease may be attributed to the concave shape of the weld pool and the increase in arc length due to the use of a beveled Y-groove and side-mounted spectral probe collection method. As the weld depth increases, the weld pool concaves, causing

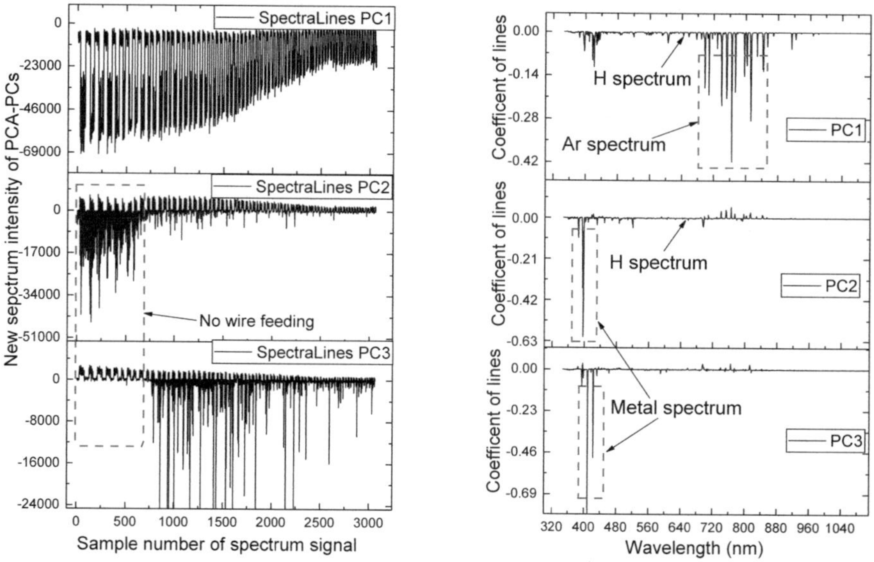

(a) The first three principal components (PCs) (b) Principal component coefficient

Fig. 2.41 PCA processing results of line spectrum

automatic contraction of the arc and potential occlusion of some arc light, leading to a decrease in the detected arc radiation intensity.

In Fig. 2.41b, the second and third principal components mainly consist of metal spectral lines. Notably, the feature curves of these components exhibit distinctly different characteristics for the same welding process in Fig. 2.41a. Particularly during the first 8 pulse cycles without wire feed, there is abundant information to be explored. Multiple weld seams with different states were subjected to PCA processing to analyze their coefficient vectors. Larger coefficient values indicate greater sensitivity to changes in the welding process state. Through comparison with the NIST database, several common spectral lines with relatively large coefficient values were identified. These findings are summarized. It is important to note that the generalization applies specifically to the welding base material and wire used in this study; results may vary for different workpiece compositions. Additionally, further comprehensive analysis is required to determine the most probable spectral line elements for the 128th and 105th pixels, considering multiple factors.

2.4.4 *Extraction of Pore Defect Features Based on PCA Coefficients*

The hydrogen-to-argon spectral intensity ratio plays a crucial role in the detection of gas porosity defects within the weld pool. Unlike other surface defects, gas porosity defects originate within the liquid weld pool, where hydrogen atoms dynamically equilibrate between the arc plasma and the weld pool. The spectral intensity of hydrogen emission lines released during this process is typically low and extremely weak, posing significant challenges for accurate porosity detection. Hydrogen emission lines directly correlate with the presence of hydrogen gas pores. By monitoring the intensity of hydrogen emission lines in the arc atmosphere, internal gas porosity defects can be indirectly characterized. However, this method is heavily influenced by factors such as the pixel accuracy of the spectrometer, integration time, acquisition method, and welding process parameters.

As observed in Fig. 2.41, the coefficients of hydrogen emission lines in the principal components are very small, indicating minimal changes in hydrogen emission line radiation during welding dynamics. Utilizing the ratio of hydrogen emission lines to argon emission lines can help mitigate the impact of non-hydrogen factors on porosity detection while enhancing sensitivity to porosity defects. Based on the PCA selection results, two different argon emission lines (Ar I 763.3 nm and Ar I 675.3 nm) were selected for ratio calculation with the hydrogen emission line (HI 656.28 nm). The PCA coefficients for Ar I 763.3 nm and Ar I 675.3 nm were 0.4182 and 0.019, respectively. Figure 2.42 illustrates the monitoring curves of the aforementioned ratio features for a complete weld seam. It can be observed that the HI 656.28 nm/Ar I 763.3 nm ratio feature curve exhibits lower noise, distinct pulse characteristics, and significant amplitude variations corresponding to porosity defects in the latter part. Therefore, higher PCA coefficients indicate better characterization capability of the constructed hydrogen ratio feature for the welding process and defects. PCA coefficient values serve as evaluation parameters for deep mining, facilitating the selection of features more sensitive to process and defect characteristics from high-dimensional data.

Hydrogen Spectral Line PCA Coefficients (PCoefHS). In this section, we present a novel method for extracting feature parameters for gas porosity monitoring based on the analysis of PCA coefficients discussed earlier. Specifically, we leverage the coefficient values associated with the hydrogen spectral lines within each principal component of PCA to characterize gas porosity defects in weld seams of varying lengths. The input sample's feature dimension corresponds to the number of spectrometer pixels, which is 1322. Meanwhile, we perform dimensionality reduction using principal component analysis, reducing the dimensionality to 20. Among the 1322 dimensions, 530 correspond to the pixels representing hydrogen spectral lines. Subsequently, we obtain a coefficient vector of size (1*20), enabling the characterization of gas porosity defects with different severities across weld seams of specific lengths. It is worth noting that two key parameters influence this algorithm. Firstly,

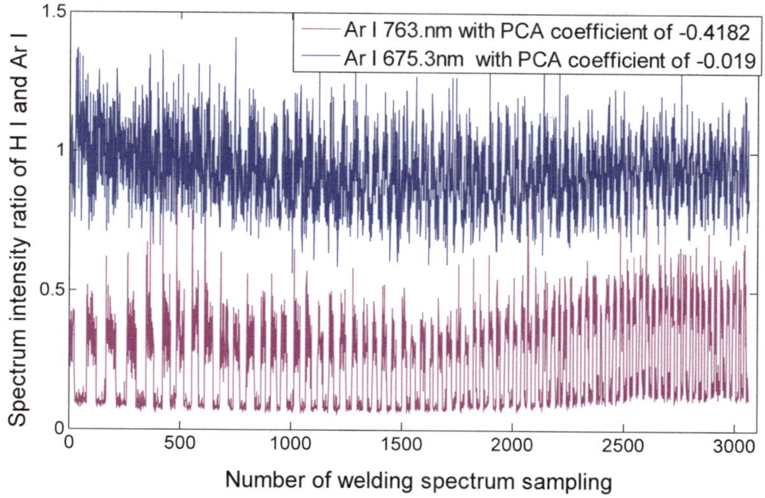

Fig. 2.42 Intensity ratio of HI (656.28 nm) to Ar I (763.3 and 675.3 nm) for different PC1 coefficients

the length of the input sample, i.e., the extent of the weld seam under examination, plays a crucial role. It should encompass various welding sample states to yield substantial coefficient values for hydrogen spectral lines. Adjusting the sample length in practical welding manufacturing can enhance the algorithm's reliability. Secondly, the reduced dimension after dimensionality reduction also impacts the algorithm's performance, albeit to a lesser extent. Careful consideration of these parameters ensures effective gas porosity detection and characterization in welding processes.

Principal Component Analysis (PCA) Statistical Features. The newly derived principal components from PCA compression can be employed for welding process detection. Based on this, we propose a method for extracting statistical features of line spectra principal components. Termed VarPCLS (Variance of Principal Components of Line Spectrum), this method involves varying the PCA input sample length from 10 to 60 and calculating the variance, kurtosis, and skewness of each principal component. As observed in the previous analysis, PC1, the first principal component, primarily consists of argon spectral lines, while the coefficients of hydrogen spectral lines are more significant in PC2 and PC3. After careful comparison, we ultimately determine that using a sample length of 30 provides optimal results, with the variance of principal components serving as the key feature parameter.

2.4.5 *Online Monitoring and Verification of Porosity Defects*

This study validates the proposed method by detecting weld seams with three different degrees of gas porosity defects. As depicted in Fig. 2.43, representative weld seams along with their X-ray inspection results are presented. For instance, Weld Seam 4 exhibits moderate gas porosity defects in the latter half due to the presence of oxide film. Similarly, Weld Seam 6, a V-groove weld overlay, also shows an oxide film in the latter half, but X-ray inspection reveals no apparent gas porosity defects. In contrast, Weld Seam 9, which has not undergone oxide film cleaning after water immersion, exhibits chain-like gas porosity defects evident in X-ray inspection. Based on these observations, three categories of gas porosity defects are defined: moderate, absent, and severe.

Firstly, traditional methods are employed to characterize and detect the three types of porosity. Utilizing the HI 656.28 nm/Ar I 763.3 nm feature, the characteristic curves depicted in Fig. 2.14 are obtained. It can be observed that the three curves show no distinct boundaries, indicating that this feature is ineffective in distinguishing between the three types of porosity defects. This could be attributed to factors such as the experimental setup and data acquisition. It suggests that traditional monitoring parameters lack sufficient robustness and reliability, necessitating further optimization to enhance the accuracy of porosity detection (Fig. 2.44).

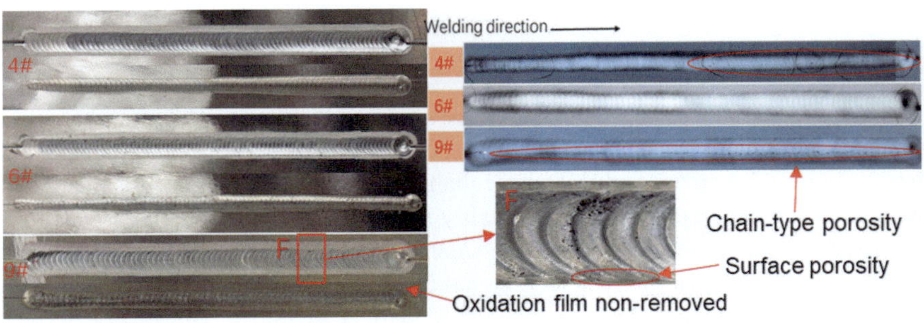

Fig. 2.43 Weld seams with varying degrees of porosity defects and X-ray inspection results

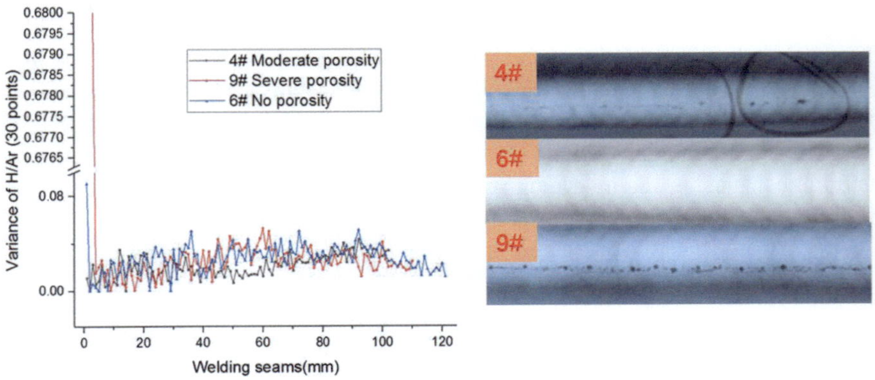

Fig. 2.44 Characteristic curves of HI (656.28 nm) and Ar I (763.3 nm) strength ratio for welds with different porosity

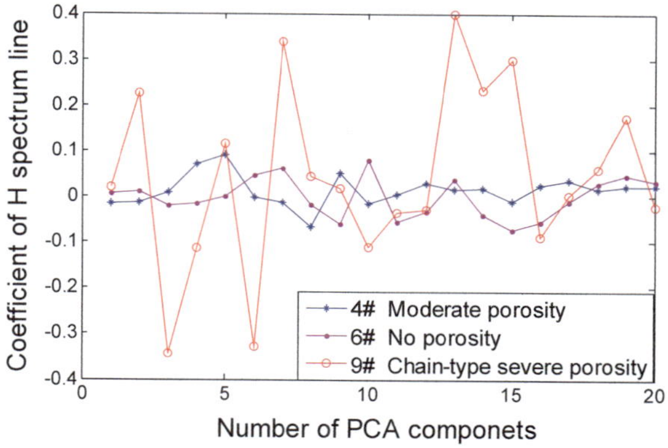

Fig. 2.45 PCoefHS characteristic curves of hydrogen spectral line principal component coefficients for different porosity welds

Further characterization and validation of the three types of gas porosity defects in weld seams are conducted using the proposed feature of hydrogen spectral line PCA coefficients (PCoefHS). The entire weld seam data is taken as input, and after PCA processing, the original 1322-dimensional spectra are reduced to 20 dimensions. From each dimension's feature vector, the coefficients corresponding to the hydrogen spectral lines are extracted, resulting in the curve graph depicted in Fig. 2.45.

From Fig. 2.45, it is evident that weld seams with different degrees of gas porosity defects exhibit varying levels of amplitude and fluctuation in the curves. Particularly, the red curve corresponding to weld seam 9# with severe gas porosity defects demonstrates pronounced fluctuations, indicating that the coefficient values corresponding to hydrogen spectral lines can to some extent reflect the severity of hydrogen gas porosity. To further quantitatively analyze the discriminative power of the feature parameters, statistical feature calculations are performed, and the results are presented in Table 2.14. It is apparent that all four statistical values of the feature curve for weld seam 9# are higher than those of the other two weld seams. This suggests that the proposed method has good identification capability for severe chain-like gas porosity. The statistical values corresponding to weld seams 4# with moderate gas porosity and 6# with no gas porosity are relatively small, with moderate discrimination capability based on the mean values. Thus, the hydrogen spectral line PCA coefficients (PCoefHS) proposed in this study can effectively detect gas porosity defects in weld seams of certain lengths. The statistical features can quantitatively assess the severity of gas porosity defects and demonstrate better robustness and identification capability than traditional hydrogen spectral line ratios.

Table 2.14 Statistical characteristics of the characteristic parameter curves in Fig. 2.37

	Max	Mean	Std	Range
4#	0.09	0.0135	0.0340	0.1573
6#	0.0796	– 0.0015	0.0439	0.153
9#	0.3978	0.0427	0.1975	0.7419

Further, we investigated the application of the t-distributed stochastic neighbor embedding (t-SNE) method for visual analysis. t-SNE, proposed by the deep learning expert Hinton, is a classic and effective nonlinear dimensionality reduction and visualization technique. This method, compared to linear methods like PCA, can effectively project data into low-dimensional space while preserving strict separation boundaries. Given the high computational complexity of this algorithm, we propose an integrated PCA-tSNE dimensionality reduction and visualization method to achieve two-stage reduction of spectral features and visualize different degrees of porosity defects. Initially, PCA is used to reduce the dimensionality of spectral samples of a certain length from 1322 to 10 dimensions. Then, the coefficients of the hydrogen spectral lines are extracted from each principal component, followed by compression to 3 dimensions using t-SNE. Figure 2.16 illustrates the monitoring curves of different degrees of porosity defects using the PCA first principal component coefficients of hydrogen spectral lines, with a spectral sample length of 60 corresponding to approximately 4 mm of weld seam. From Fig. 2.46a, distinct pulse characteristics induced by the pulsed current can be observed, and there are differences in the amplitude distribution among the three types of pores. Figure 2.46b further demonstrates the pulse characteristics, but the sample distribution is not sufficiently clear. The results of the second-stage dimensionality reduction by t-SNE on the aforementioned features are shown in Fig. 2.47. Figure 2.47a displays the ability of the t-SNE-compressed three-dimensional features to identify different pores, with the first two dimensions showing better performance. Figure 2.47b further illustrates their distribution in the two-dimensional feature space, where the three types of pores can be clearly distinguished.

2.4.6 Conclusion

This study investigates and establishes a strong correlation between porosity and the hydrogen spectral line (656.28 nm) using principal component analysis (PCA). Based on this correlation, a novel online porosity monitoring method is proposed, leveraging the principal component coefficients of the hydrogen spectral line (PCoefHS). Finally, different degrees of porosity defects are visualized through integrated PCA-tSNE (t-distributed Stochastic Neighbor Embedding) technique. The following conclusions can be drawn:

(1) By intentionally preserving surface oxide films, various degrees of internal porosity defects were fabricated. Through multiple microscopic characterization methods, it was observed that two types of internal porosity, hydrogen porosity and magnesium vapor porosity, can be generated during aluminum alloy pulsed Gas Tungsten Arc Welding (GTAW) process. Magnesium plays a significant role in the formation and growth of these pores during the dynamic changes in the weld pool. The average magnesium content inside the magnesium vapor pores is twice that of the outer edge.

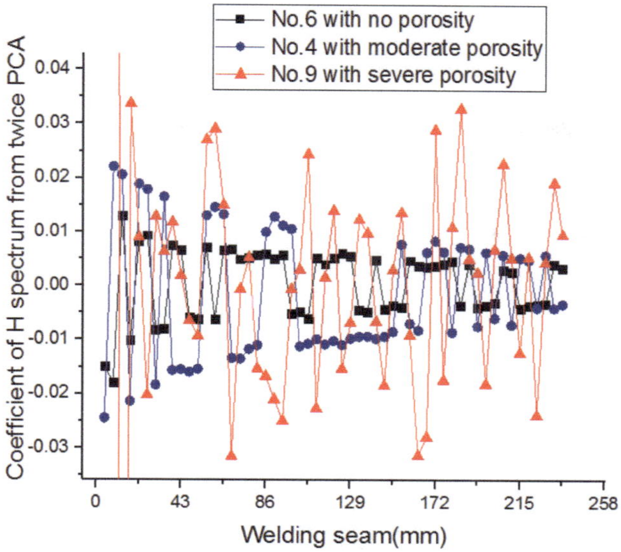

(a)Characteristic curve of the first principal component of hydrogen spectral lines.

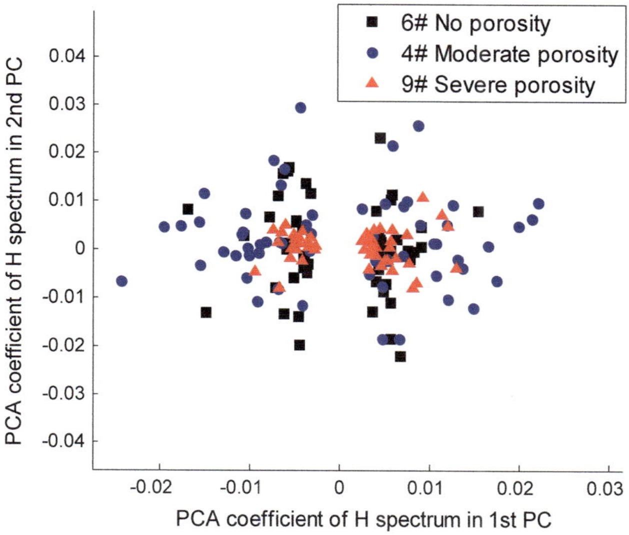

(b) Characteristic distribution of the first and second principal components.

Fig. 2.46 Characteristic curves and spatial distribution of different porosity welds

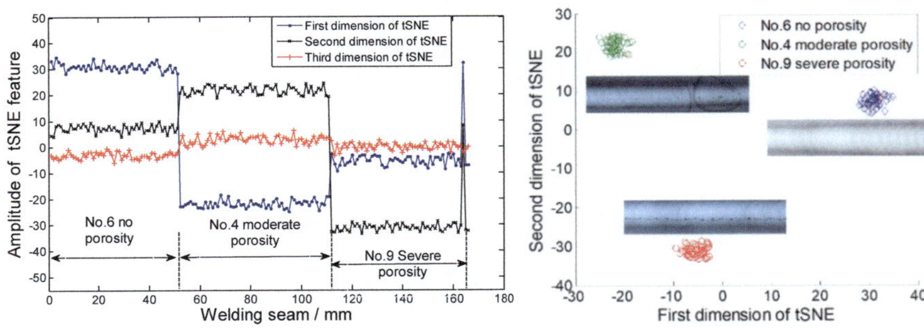

(a) 1-D visualization (b) 2-D visualization

Fig. 2.47 PCA-Tsne visualization of welds with different degrees of porosity

(2) A strong correlation was established between the PCA coefficient vectors of hydrogen spectral lines and internal porosity. These coefficients serve as effective tools for deep mining, allowing for the quantitative selection of feature spectral lines highly correlated with welding quality.
(3) An integrated PCA-tSNE spectral dimensionality reduction and visualization technique was developed. Validation results demonstrate its effectiveness in detecting and distinguishing severe, moderate, and no-porosity weld seams in aluminum alloy pulsed GTAW processes.

References

1. Zhiyong Li, Binjing Ding, Huan Li et al (2009). MIG Welding Quality Diagnose Based on Arc Light Spectrum [J]. Journal of mechanical engineering, 45(04):197–202.
2. Junyue Li, Yonglun Song, Huan Li et al (2002). The basic theory and methods of welding arc spectral information [J]. Transactions of the china welding institution, 23(06):5–8.
3. Mirapeix J, Vila E, Valdiande J J et al (2016). Real-time detection of the aluminium contribution during laser welding of Usibor1500 tailor-welded blanks[J]. Journal of Materials Processing Technology, 235:106–113.
4. Song L, Huang W, Han X et al (2017). Real-Time Composition Monitoring Using Support Vector Regression of Laser-Induced Plasma for Laser Additive Manufacturing[J]. IEEE Transactions on Industrial Electronics, 64(1):633–642.
5. Zhang Z, Yu H, Lv N et al (2013). Real-time defect detection in pulsed GTAW of Al alloys through on-line spectroscopy[J]. Journal of Materials Processing Technology, 213(7):1146–1156.
6. Huang Y, Wu D, Lv N et al (2017). Investigation of porosity in pulsed GTAW of aluminum alloys based on spectral and X-ray image analyses[J]. Journal of Materials Processing Technology, 243:365–373.
7. Yu H, Xu Y, Song J et al (2015). On-line monitor of hydrogen porosity based on arc spectral information in Al–Mg alloy pulsed gas tungsten arc welding[J]. Optics & Laser Technology, 70:30–38.
8. Huang Y, Wu D, Zhang Z et al (2017). EMD-based pulsed TIG welding process porosity defect detection and defect diagnosis using GA-SVM[J]. Journal of Materials Processing Technology, 239:92–102.

Chapter 3
Optical Spectroscopy-Based Multiple Defects Detection for GTAW

3.1 Random Forest-Based Defect Detection for GTAW Using Optical Spectroscopy

3.1.1 Welding Defect Design

The welding utilized 5A06 aluminum alloy, measuring 300 mm in length, 50 mm in width, and 4 mm in thickness. Employed in the experiment was a pulse square wave alternating current for welding. The peak period and base period were set at 0.5 s each, with a peak current of 240 A and a base current of 50 A. The current pulse frequency was 1 Hz, while the AC frequency stood at 50 Hz, referencing Fig. 3.1a. The experimental setup comprised a robotic welding system, a process-controlling system, and a multiple signal acquisition system (as depicted in Fig. 3.1a). A Yaskawa Industry robot oversaw the welding process and its running speed. Developed using LabVIEW application software, the system enabled the triggering, acquisition, and storage of all parameters and collected signals. It also facilitated control over welding current, wire feed speed, data collection time, and sample frequency. For real-time arc spectrum emission signal collection, an AvaSpec-1350F-USB2 spectrometer, consisting of a CCD spectrometer, optical probe, and fiber, was employed. This spectrometer covered a wavelength range from 350 to 1100 nm, with 1,350 pixels and a minimum integration time of 0.7 ms.

To ensure the spectrometer probe remains relatively stationary during the arc welding process, it is affixed to the welding robot arm and aligned precisely with the arc center. To moderate the acquisition process effectively, a full-wavelength dimming plate with a 10% transmittance is positioned in front of the probe. The distance between the probe and the arc center is maintained at 21 cm, with a spectral integration time of 1.5 ms. At the onset of welding on the plate, spot welding is executed for 8 s. The welding speed is set at 16 cm/min, with the arc length maintained at 3 mm. Argon flow rate is controlled at 15 L/min, and wire feed speed at 12 mm/s.

Z. Zhang et al., *Multiple Sensing-Driven Feature Extraction and Defect Detection*,
https://doi.org/10.1007/978-981-96-4808-5_3

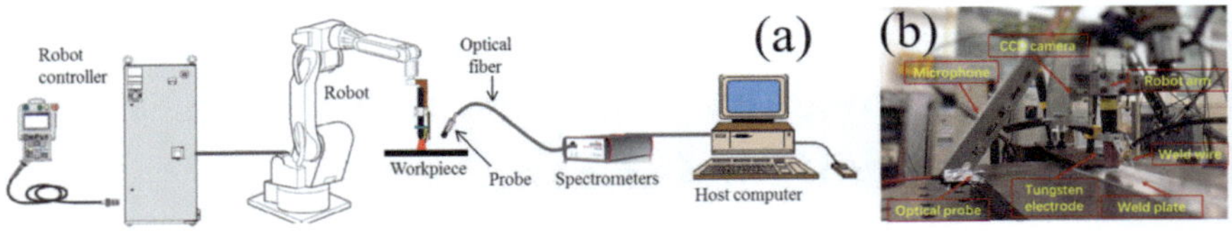

Fig. 3.1 Experiment system for robotic welding. **a** Experiment system for robotic welding, **b** multiple sensory monitoring system

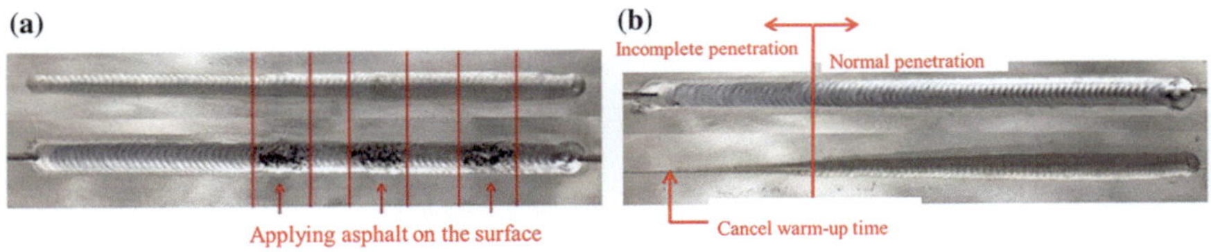

Fig. 3.2 Welding defect specimen. **a** Porosity defect, **b** incomplete penetration

Varied welding parameters are adjusted as per different experimental conditions. Upon determining optimal welding parameters, defects such as incomplete penetration are induced by eliminating preheating of the base material. Additionally, a burn-through defect test piece is created by reducing the local thickness of the base material. In this study, porosity defects are simulated by applying asphalt at fixed positions of the weld, with a focus on qualitative analysis of severe porosity defects. Subsequent research will address minor defects. The front and back welds of the defective test piece obtained from the experiment are illustrated in Fig. 3.2.

3.1.2 Preprocessing of Arc Spectrum

The preprocessing of arc spectrum consists of three main stages: identifying element and line spectra, removing background spectra, and eliminating pulse interference. A novel approach, combining Step searching and Cubic Spline Interpolation (SsCSI), was introduced for effectively separating background spectra from line spectra. Subsequently, a threshold method was employed to eliminate pulse interference. Ultimately, it was observed that the spectrum feature in the peak current exhibited heightened sensitivity to weld defects, thus emphasizing its significance in the preprocessing phase.

The lines spectrum emitted from the arc plume plays a critical role in assessing welding stability and quality. These spectra can be categorized into three main classes: hydrogen lines spectrum, argon lines spectrum, and metal lines spectrum. Through a comparative analysis of experimental data with atomic radiation spectrum data

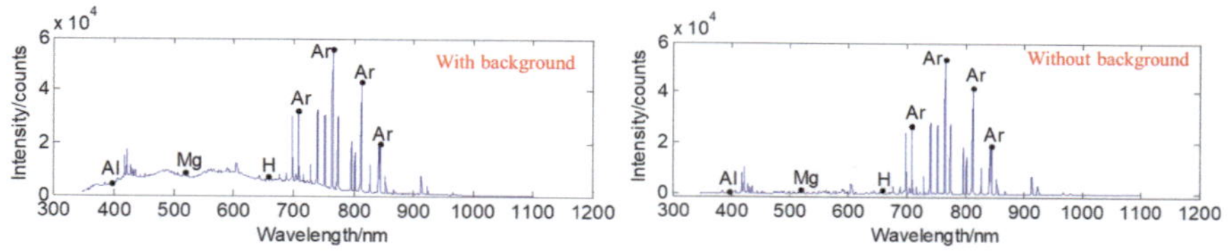

Fig. 3.3 Lines identification and background spectrum removing

sourced from the National Institute of Standards and Technology (NIST), we identified 20 distinct argon lines spectra, 1 hydrogen lines spectrum, and 4 metal lines spectra. These discernible lines have been clearly annotated in Fig. 3.3.

The arc spectrum is composed of a combination of continuous and line spectra at any given moment. Typically, the continuous spectrum is regarded as background noise and thus requires removal, leaving the line spectrum to be isolated. In this study, a novel method for separating the background spectrum from the line spectrum was introduced, leveraging Step searching and Cubic Spline Interpolation (SsCSI), as illustrated in Fig. 3.4. Initially, local minimum values were identified through step searching, followed by cubic spline interpolation to derive the lower envelope signal of the spectral data. By subtracting the background spectrum obtained through this process from the original spectral signal, the line spectrum signal was isolated, as depicted in Fig. 3.4.

Another essential preprocessing step involves eliminating pulse interference induced by the pulse alternating current. During peak current, the intensity of arc light and spectrum emission is significantly higher compared to base current. Consequently, the spectral signal intensity at peak current exceeds that at base current, resulting in periodic variations in the spectral signal at the same frequency as the current. Figure 3.5 displays the acquired arc current and spectral signal, illustrating this phenomenon.

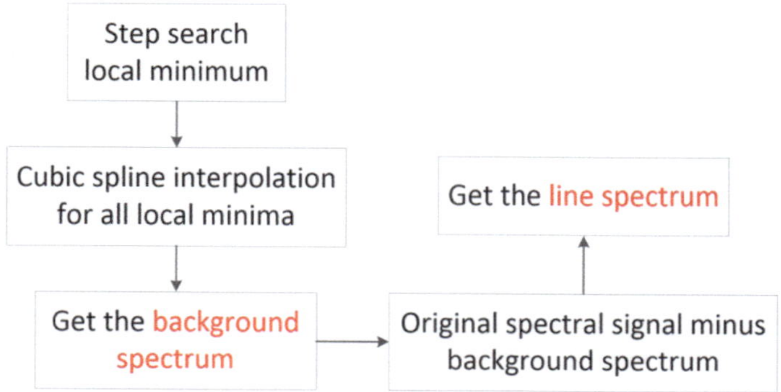

Fig. 3.4 Background spectrum removal based on step searching and cubic spline interpolation (SsCSI)

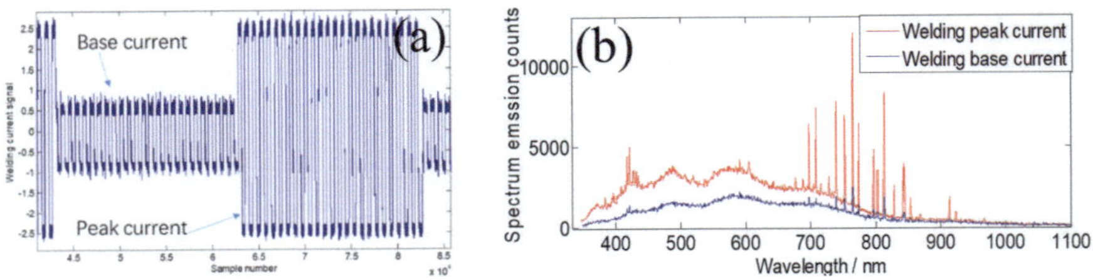

Fig. 3.5 Acquired welding process signal. **a** Pulsed AC-square wave welding current signal, **b** arc spectrum in peak and base current

To mitigate pulse interference and assess the sensitivity of both base and peak values to defects, the threshold method was employed to differentiate between base and peak spectrum signals. When three consecutive points surpass the threshold, it is indicative of line spectrum during peak current. Conversely, when three consecutive points fall below the threshold, it signifies line spectrum during base current. Mean values were then calculated for both peak and base states within each pulse period.

Figure 3.6 illustrates the original feature curve based on Ar I (706.72 nm) line spectrum during the welding process, while Fig. 3.7 depicts the results of pulse removal. It is evident from Fig. 3.7 that pulse interference has been effectively eliminated from both peak and base spectrum signals. Notably, the sensitivity of base and peak features exhibits distinct patterns in response to the same welding defects. Specifically, three instances of burn-through defects are marked in Fig. 3.6. However, in Fig. 3.7, the peak feature of the spectrum demonstrates corresponding changes, whereas the base spectrum feature exhibits minimal changes when defects occur. Consequently, the spectrum feature during peak current has proven to possess superior sensitivity to weld defects. To streamline calculations and enhance online efficiency, spectrum signals during peak current were retained for further defects detection.

3.1.3 Feature Extraction of Arc Spectrum

During the welding process, various random factors such as probe movement and heat accumulation can influence the intensity of acquired spectra, resulting in inconsistent intensity even under the same welding parameters. To enhance the robustness of the features, this paper proposes a dimensionless index based on different spectral intensity ratios rather than absolute spectrum intensity.

A comprehensive knowledge base of spectrum features for weld defects is established, encompassing 19 Ar I line spectrum ratios, 4 metal line spectra, and 1 hydrogen line spectrum. As depicted in Fig. 3.7, the spectrum feature during peak current exhibits superior sensitivity to weld defects. To streamline calculations and enhance online efficiency, spectrum signals during peak current are preserved for further defects detection. Thus, the listed 25 features in Table 3.1 are augmented to 50 features, incorporating two types of statistics for each ratio feature. The flowchart

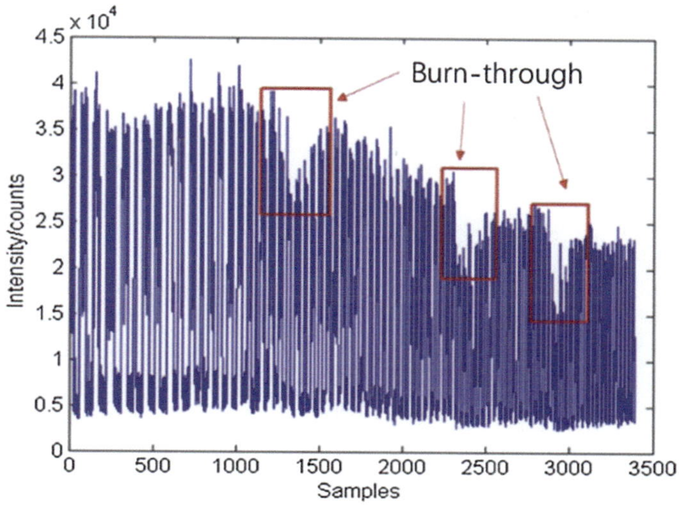

Fig. 3.6 Original spectral signal of Ar I line spectrum (706.72 nm)

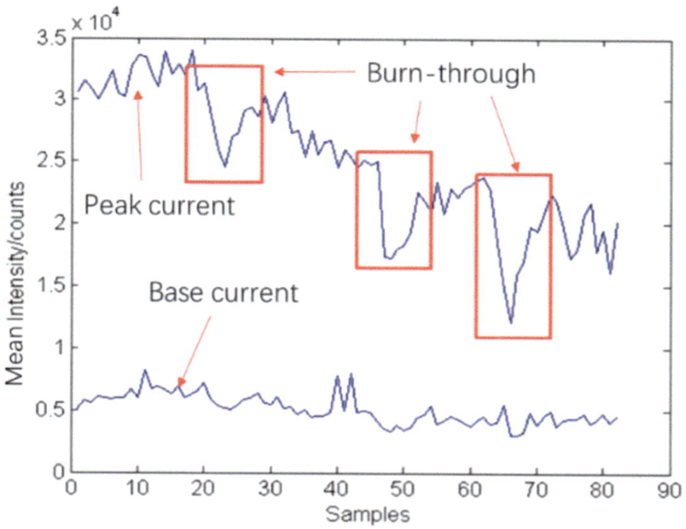

Fig. 3.7 Base and peak feature curves of Ar I line spectrum (706.72 nm) after pulse removing

of the feature extraction process is illustrated in Fig. 3.8, while the spectrum ratio features are enumerated in Table 3.1. Subsequently, the sensitivity of these ratio features to various weld defects is meticulously analyzed.

During the welding of aluminum alloy, liquid aluminum has the capability to absorb a significant amount of hydrogen while in a liquid state, but hardly any when solidified. Consequently, the majority of porosity generated during aluminum

Table 3.1 Lines spectrum and its ratio feature

Number	Ratio feature meaning	Number	Ratio feature meaning
1	Mg (383.83 nm) to background spectrum	14	Ar (727.29 nm) and Ar (714.70 nm) ratio
2	Al (394.61 nm) to background spectrum	15	Ar (751.46 nm) and Ar (727.29 nm) ratio
3	Al (396.25 nm) to background spectrum	16	Ar (763.51 nm) and Ar (751.46 nm) ratio
4	Mg (517.55 nm) to background spectrum	17	Ar (763.51 nm) and Ar (794.81 nm) ratio
5	H (656.28 nm) to background spectrum	18	Ar (810.36 nm) and Ar (794.81 nm) ratio
6	Ar (603.21 nm) and Ar (641.63 nm) ratio	19	Ar (811.53 nm) and Ar (810.36 nm) ratio
7	Ar (675.28 nm) and Ar (641.63 nm) ratio	20	Ar (811.53 nm) and Ar (826.45 nm) ratio
8	Ar (687.12 nm) and Ar (675.28 nm) ratio	21	Ar (840.82 nm) and Ar (826.45 nm) ratio
9	Ar (687.12 nm) and Ar (693.76 nm) ratio	22	Ar (842.46 nm) and Ar (840.82 nm) ratio
10	Ar (696.54 nm) and Ar (693.76 nm) ratio	23	Ar (842.46 nm) and Ar (852.14 nm) ratio
11	Ar (696.54 nm) and Ar (703.02 nm) ratio	24	Ar (852.14 nm) and Ar (866.79 nm) ratio
12	Ar (706.72 nm) and Ar (703.02 nm) ratio	25	Ar (675.28 nm) and H (656.28 nm) ratio
13	Ar (706.72 nm) and Ar (714.70 nm) ratio	–	–

alloy welding is hydrogen-induced porosity. The formation of porosity is a dynamic process influenced by the balance between gravity, flow resistance of liquid metal, and metal vapor. Hydrogen, during its escape process, may carry some metal ions, leading to an increase in the concentration of metal ions in the arc range and consequently an increase in line intensity.

The left image of Fig. 3.9 illustrates the peak mean curves of the original spectral intensity for different line spectra. It can be observed that H I line, Al I line, and Mg I line spectra exhibit varying degrees of increase when surface porosity defects occur. Notably, the Al I line spectrum displays the most significant change, potentially attributed to its higher concentration and weight in the material composition. Furthermore, the downward trend observed in the original feature curves has been effectively mitigated through the utilization of ratio features.

The arc length may vary for different penetration states, leading to differences in the intensity of the Ar I line spectrum. Consequently, the intensity changes of the Ar I line spectrum can serve as an effective indicator of arc length variation. To

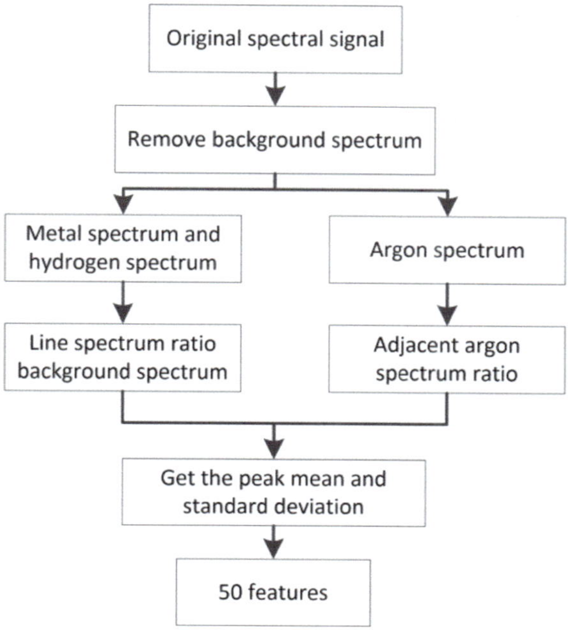

Fig. 3.8 The flowchart of feature extraction

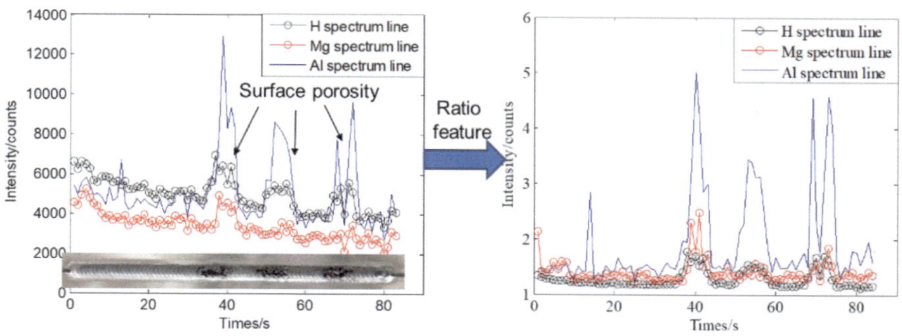

Fig. 3.9 Original feature curve and improved ratio feature for porosity defect

capture this relationship, the ratio feature of adjacent argon spectra is calculated and presented in Fig. 3.10. It is evident from the Fig. 3.10 that the ratio feature effectively eliminates downward trends and enhances the significance of defect characteristics.

By calculating the mean value of line intensity, we obtained 19 ratios for argon spectra, 4 ratios for metal spectra, and 1 ratio for hydrogen spectra, all compared to background spectra. Additionally, the ratio of hydrogen spectra to argon spectra from literature data was extracted. This yields a total of 25 dimensions. However, signals with the same mean value may exhibit different fluctuations. To provide a

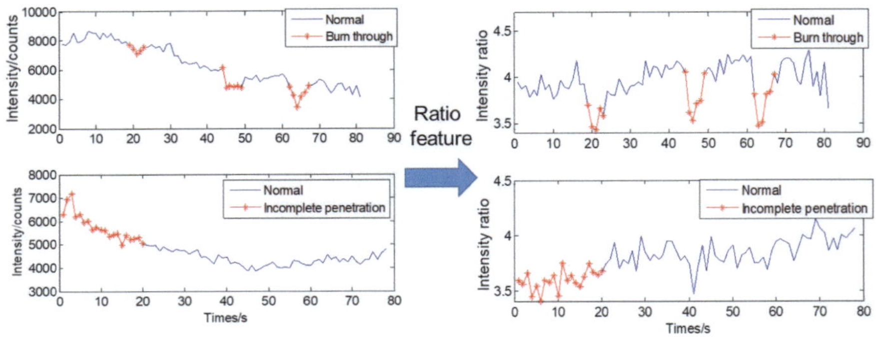

Fig. 3.10 Ar I line spectrum ratio (714.70 nm/706.72 nm) for different welding defects

more comprehensive description of signal fluctuations, we also calculated the peak mean and standard deviation of the intensity ratio for each line. As a result, this paper extracted a total of 50 features related to line intensity ratios.

3.1.4 Spectral Feature Selection Based on Random Forest

In this section, we introduce a novel quantitative index of feature importance, leveraging the concepts of MDA and MDG discussed in Sect. 3.2. This index facilitates the reduction of the original feature space and effectively eliminates redundancy. By employing this index along with Random Forest (RF), we identify an optimal feature subset comprising 6 ratio features. MDA and MDG assess feature importance based on classification accuracy and Gini index, respectively. To integrate their evaluation outcomes, we propose a comprehensive index. Initially, the two feature importance indices are normalized, following which the mean values of the two indices are computed. The formula for this comprehensive index is as follows:

$$J_c(i) = \left(\sum_{k=a,b} \frac{\max(J_k) - J_k(i)}{\max(J_k) - \min(J_k)} \right) / 2 \tag{3.1}$$

where J_a is the value of MDA and J_b is the value of MDG, which can be obtained based on formulas 3.3 and 3.4 in Sect. 3.2. In addition, MDA and MDG are calculated inside the RF classification model, and are the key to achieve model optimization by randomly constructing feature subsets and iterations.

To establish the RF classification model, we utilized a dataset comprising 190 samples, including 43 normal, 43 burn-through defects, 44 porosity defects, and 60 incomplete penetration defects. To assess the algorithm's stability and adaptability to diverse scenarios, we randomly selected 150 samples for training in each experiment, reserving the remaining 40 samples for testing.

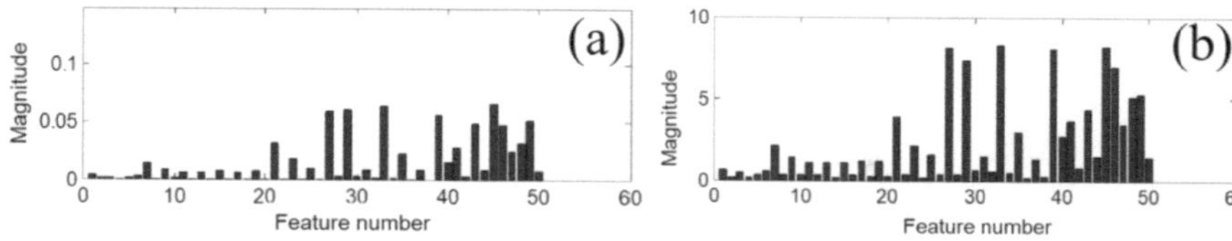

Fig. 3.11 Importance index diagrams for 50 spectrum features. **a** MDA results based on RF, **b** MDG results based on RF

Generally, a larger forest scale contributes to a more stable model. To ensure model stability, we set the random forest scale to 500 and input the samples into the model. The Mean Decrease Accuracy (MDA) and Mean Decrease Gini (MDG) results for each spectrum feature are listed in Fig. 3.11.

Tables 3.2 and 3.3 provide the physical meaning of the top 10 features based on the Mean Decrease Accuracy (MDA) and Mean Decrease Gini (MDG) feature importance rankings, respectively. It is observed that the ratio features of Al I (394.61 nm) line spectrum and H I (656.28 nm) line spectrum exhibit the highest importance. Additionally, Mg I (383.83 nm), Al I (396.25 nm), and several Ar I line spectra also appear among the top 50 features.

Based on the importance index derived from MDA and MDG, the features are sorted in descending order, and the first n (n = 1,…, 50) features are inputted into the random forest model for defects recognition. The classification accuracy of the varying features is depicted in Fig. 3.12, revealing an average accuracy rate of 97.75% with the input of the six most effective features. As more features are inputted, the recognition accuracy fluctuates, indicating that the first six features may suffice to achieve the correct classification rate. However, as additional features are added, some less effective features may decrease the recognition accuracy of the model. Nevertheless, increasing the number of features enhances the model's stability while

Table 3.2 Feature importance ranking based on MDA

Sequence	Physical meaning of the feature	Value of MDA
1	Mean value of Al (394.61 nm) to background spectrum	0.0729
2	Mean value of H (656.28 nm) to background spectrum	0.0558
3	Mean value of Ar (811.53 nm) and Ar (810.36 nm) ratio	0.0519
4	Mean value of Mg (383.83 nm) to background spectrum	0.047
5	Mean value of Mg (517.55 nm) to background spectrum	0.0479
6	Mean value of Ar (811.53 nm) and Ar (826.45 nm) ratio	0.0474
7	Mean value of Ar (842.46 nm) and Ar (840.82 nm) ratio	0.0464
8	Mean value of Ar (763.51 nm) and Ar (751.46 nm) ratio	0.0375
9	Mean value of Ar (675.28 nm) and H (656.28 nm) ratio	0.0371
10	Mean value of Al (396.25 nm) to background spectrum	0.0334

Table 3.3 Feature importance ranking based on MDG

Sequence	Physical meaning of the feature	Value of MDG
1	Mean value of Al (394.61 nm) to background spectrum	8.5677
2	Mean value of H (656.28 nm) to background spectrum	7.6976
3	Mean value of Ar (811.53 nm) and Ar (810.36 nm) ratio	7.0324
4	Mean value of Ar (811.53 nm) and Ar (826.45 nm) ratio	6.9484
5	Mean value of Ar (842.46 nm) and Ar (840.82 nm) ratio	6.5380
6	Mean value of Ar (675.28 nm) and H (656.28 nm) ratio	5.3909
7	Standard deviation of Al (396.25 nm) to background spectrum	5.2464
8	Mean value of Mg (517.55 nm) to background spectrum	4.7628
9	Standard deviation of Al (394.61 nm) to background spectrum	4.7495
10	Mean value of Ar (763.51 nm) and Ar (751.46 nm) ratio	4.7175

also increasing feature redundancy. From the above analysis, it is evident that the most important six features essentially encompass the most effective features for identifying different welding conditions.

Furthermore, Table 3.4 presents a comparison between the recognition accuracy and calculation speed of 6-dimensional prominent features and all 50-dimensional features. The experimental computer environment is configured with an Intel Core i3-4150 CPU @ 3.50 GHz, running Windows 8.1, and utilizing MATLAB R2014a as the programming environment. It is observed that the calculation speed of 6-dimensional prominent features is increased by 72.52% while the accuracy of 6-dimensional

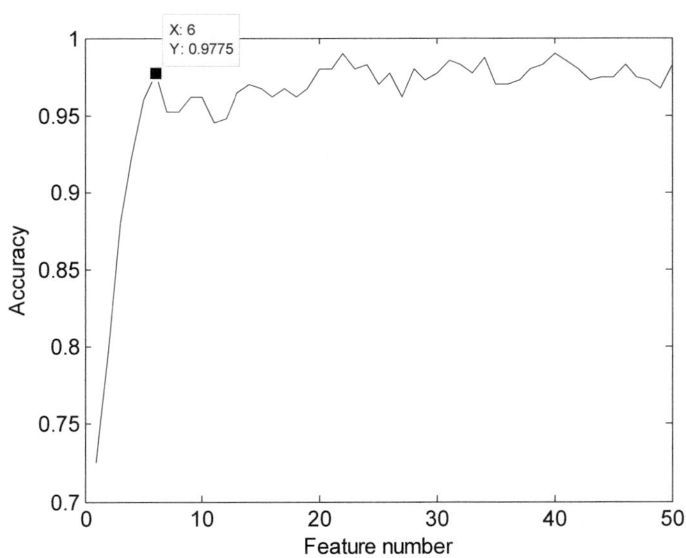

Fig. 3.12 Model accuracy of different feature numbers

Table 3.4 Recognition results of different input feature

Number of input feature	Average accuracy rate	Time
6	97.75	0.0739 s
50	98.15	0.2689 s
Change rate	– 1.03%	72.52%

prominent features remains almost unchanged compared to using all features. This enhancement in calculation speed can effectively improve the efficiency of welding defect recognition.

The physical meaning of the top 6 effective feature rankings is displayed in Table 3.5, along with the calculated index values of feature importance obtained by combining the Mean Decrease Accuracy (MDA) and Mean Decrease Gini (MDG) metrics as calculated in Eq. 4.1. Given the normalization performed, the feature importance values range from 0 to 1. It is evident that the 6 effective features encompass the ratio of metal spectral intensity, the ratio of hydrogen spectral intensity to background spectral intensity, and the ratio of different argon spectral intensities. Notably, the selected effective features represent the peak mean characteristics of line intensity, suggesting that mean line intensity is more indicative of changes in weld state compared to standard deviation.

Based on the optimal feature subset results, construction pattern of the ratio feature and selection laws of line spectrum were carefully analyzed, which can be helpful for the selection of spectral lines or any other monitoring signals with high dimension.

The extraction of Ar lines spectrum ratios at the outset of feature extraction resulted in 19 ratio features initially. However, after feature selection and reduction, only 3 ratio features of Ar spectrum were retained among the 6-dimensional salient features, involving 5 Ar I lines spectra, as depicted in Fig. 3.13. The observed patterns of these 3 ratio features of Ar spectrum are summarized as follows:

Table 3.5 Selected feature subset with 6 features and its physical meaning based on MDA and MDG

Importance ranking	Physical meaning of feature	Proposed index value
1	Mean value of Al (394.61 nm) to background spectrum	1
2	Mean value of H (656.28 nm) to background spectrum	0.8317
3	Mean value of Ar (811.53 nm) and Ar (810.36 nm) ratio	0.7614
4	Mean value of Ar (842.46 nm) and Ar (840.82 nm) ratio	0.7289
5	Mean value of Ar (811.53 nm) and Ar (826.45 nm) ratio	0.7224
6	Mean value of Mg (383.83 nm) to background spectrum	0.6066

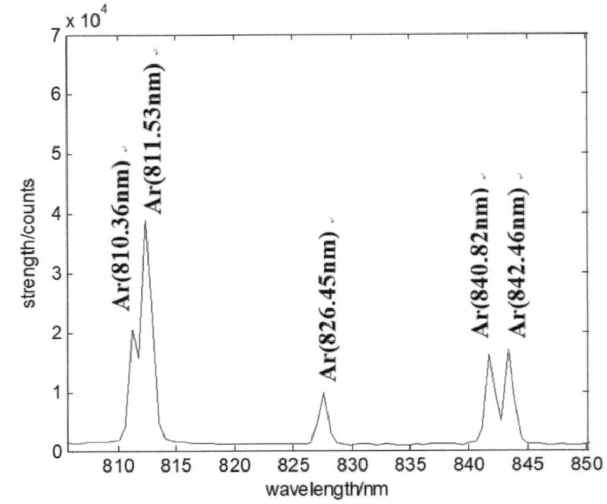

Fig. 3.13 Selected Ar I lines spectrum

(1) The wavelength difference between the two argon spectra is small, indicating minimal changes in the response characteristic curve of the spectrometer. This suggests that the spectral signal characteristics remain similar during analog-to-digital conversion, contributing to consistent spectral data.
(2) Background spectral intensities are small and stable. Given the complexity of the formation mechanism of arc signals in the welding process, the stability and low intensity of background spectra imply that signals in this wavelength band primarily consist of excited radiation, minimizing nonlinear interference from other radiation sources.
(3) Atomic spectral radiance is high, with clear spectral lines for elements and minimal interference from neighboring elements. This ensures accurate readings of spectral line intensities with minimal errors.

The summarized characteristics of argon spectra can provide valuable insights for the selection of spectral lines for various elements in future applications. Understanding these patterns can aid researchers in effectively choosing spectral lines for analysis in diverse contexts.

3.1.5 Recognition of Welding Defects Based on Random Forests

In order to validate both the proposed feature importance index and the RF defects classification model, a comparative analysis was conducted among three different methodologies: RF, Radial Basis Function (RBF) neural network, and BP neural network. As shown in Table 3.6, for the 6 significant features characterized by varying importance indices, RF demonstrated superior performance in terms of

Table 3.6 Comparison with different models and feature subset

Input features	Model	Highest correct rate (%)	Minimum correct rate (%)	Average correct rate (%)	Variance
The most important 6 features	RF	100	90	97.75	0.0001
	RBF	65	20	43.55	0.0098
	BP	97.5	52.5	83.7	0.0128
The least important 6 features	RF	60	35	49.63	0.0058
	RBF	40	17.5	28.75	0.0037
	BP	52.5	10	31.9	0.0073

average correctness rate at 97.75%, with the lowest variance compared to the RBF and BP classification models. Furthermore, the correctness rate notably decreased for the least important 6 features across all models—RF, RBF, and BP. These results not only emphasize the effectiveness of the proposed feature importance index but also highlight the superior performance of the weld defects classification model based on RF.

3.1.6 Conclusions

This paper researched the on-line defects detection method for aluminum alloy in robotic arc welding based on random forest and arc optical spectrum. The conclusion can be drawn as follows:

1. It was found that the spectrum feature in peak current has demonstrated better sensitivity to weld defects than base current.
2. Al I (394.61 nm) line spectrum and H I (656.28 nm) line spectrum have shown the highest importance following by Mg I (383.83 nm), Al I (396.25 nm) and several Ar I line spectrum among 50 features.
3. Comparing with RBF and BP classification models, RF has shown the best performance in identifying three typical defects including incomplete penetration, burn-through and porosity with the average correct rate of 97.75% and the smallest variance.

3.2 Advanced Spectral Features-Based Wire Feed Monitoring for GTAW

Monitoring the process state of aluminum alloy welding is of vital importance for ensuring the quality stability of aerospace components and promoting the intelligent manufacturing of robot welding. Research has been conducted to explore the fuzzy correlation between arc spectroscopy and defects, as the existing methods for knowledge extraction from arc spectroscopy are limited. Through experiments, a sensitive zone for probe positioning has been determined to ensure the reliability of data collection. The principal component analysis method has been employed to select metal spectral lines such as AlI, MgI, and FeI, enabling quantitative evaluation of the sensitivity of each spectral line to changes in state. Based on the characteristic coefficients of metal spectral line principal components, the dynamic response patterns of different elements have been analyzed, revealing a strong correlation between FeI spectral line and wire feeding status. Wire feeding status detection has been carried out based on SPC statistical process control, and through repeated validation in various welding experiments, the results indicate that this method exhibits high stability and strong anti-interference capability.

3.2.1 Welding Test System

The experimental system consists of a robotic welding system, a multi-information synchronous data acquisition system, and a process control system. The system utilizes LabVIEW software and data acquisition cards to control welding current, wire feed speed, and welding speed. It synchronously triggers multiple signal sensors to achieve synchronous acquisition and storage of current, voltage, sound, images, and arc spectra. Y-groove filling and welding of 4 mm thick 5A06 aluminum alloy were performed, with specific welding parameters listed in Table 3.7 the arc spectrum was collected in real-time using a fiber optic spectrometer (AvaSpec-1350F-USB2) fixed at the end of the robot. The collection range is 350 ~ 1100 nm, with a camera pixel of 1350, a set integration time of 1.5 ms, and a sampling rate of approximately 70 Hz. The spectrometer is highly sensitive to the probe position. Before the experiment, the probe was moved, and the spectrum was collected in real-time to analyze and select a reasonable probe position. As shown in Fig. 3.14, the probe was fixed in relation to the arc, and the probe was slowly adjusted from top to bottom and from left to right. Figure 3.15 shows the intensity curve of the magnesium elemental line spectrum during the movement process. It can be seen from the Fig. 3.15. that when the probe moves up and down, the spectral intensity reaches its maximum near the arc, and the arc light may be too strong and may exceed the maximum range. When the probe is slightly above or below, the intensity of MgI drops sharply, indicating that the tungsten electrode blocks the arc light, and the probe is sensitive to vertical movement. It is best to fix it in an area where data can be collected steadily, that is, in a stable and changing area. In comparison, the intensity changes from left to right is not so drastic because there are no obvious obstructions. It is suggested to collect data in the region with a large and stable peak value. Finally, the spectrum probe in Fig. 3.14b was used to face the center of the arc, and a 10% light reducing sheet was added to make the spectral intensity near 2/3 of the range.

Table 3.7 Welding experiment parameters

Parameter	Value	Parameter	Value
Impulse frequency/Hz	1	Welding speed/cm/min	16
Peak current/A	240	Wire feed rate/mm/s	12
Base current/A	50	Tungsten diameter/mm	3.2
Duty cycle/%	50	Welding wire type/diameter/mm	SG-AlMg6/1.6
Argon flow/L/min	15	Sheet size/mm	300 × 50 × 4

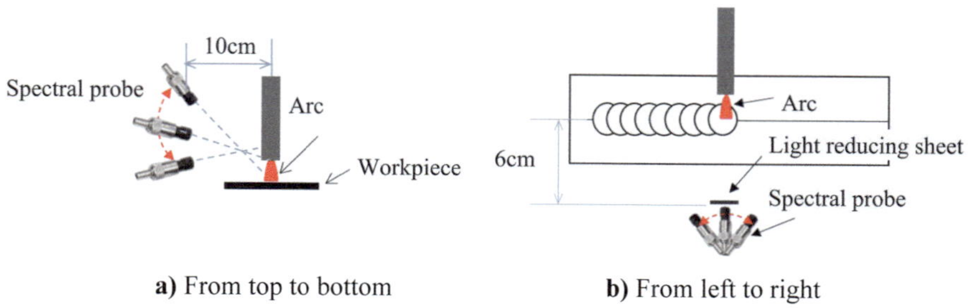

Fig. 3.14 Position adjusting of spectrum probe

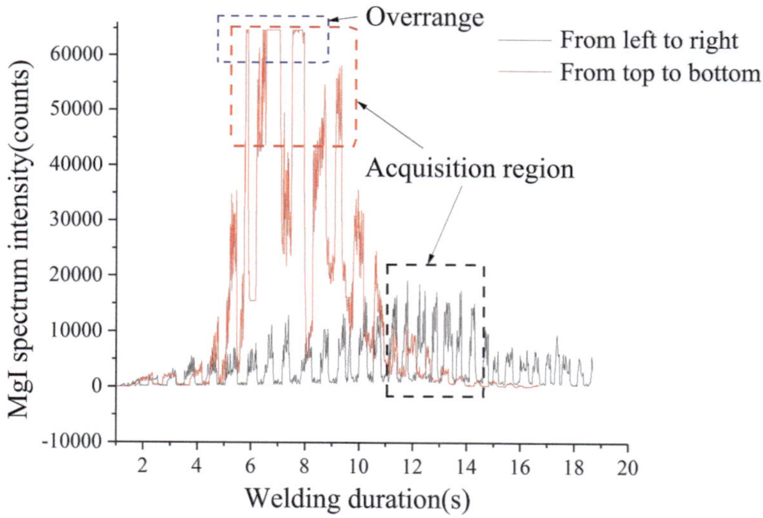

Fig. 3.15 Mg I Spectrum curves during moving the probe

3.2.2 Metal Spectrum Line Selection

The spectral information in the metal spectrum segment was deeply explored using Principal Component Analysis (PCA) to identify metal spectral line elements that are strongly correlated with welding dynamic changes and have low coupling. The full spectrum signals of arc spectra collected at the peak and base values of welding current are obtained, as shown in Fig. 3.16 Observing the change patterns of spectral lines, it is found that the metal spectral lines do not completely disappear or reappear due to different welding states; their radiation intensity is dynamically changing. This may be related to the method of integration and accumulation of data acquisition. It also indicates that using the absolute value of spectral intensity is not effective in characterizing such changes, and statistical probability methods may be more suitable for characterizing such spectral line changes. PCA can extract linear patterns in the data, and the calculated eigenvectors can be used to quantitatively characterize the sensitivity or importance of each feature, making it suitable for analyzing this type of data. Finally, the spectral segment of 382.2 ~ 429.64 nm, with a total of 81 dimensions, was selected for PCA spectral line selection and weight analysis.

The method of removing the envelope was used to separate the background spectrum from the line spectrum of the spectral signal, and PCA processing was performed separately. Figure 3.17 shows the separated line spectrum and background spectrum, along with the change in PCA contribution rate with the increase of the number of main component features. The input data consisted of 3638 spectral samples from a weld seam, including four welding states: no wire feed, wire feed, wire feed with grinding, and wire feed without grinding. From Fig. 3.17, it can be seen that the contribution rate curves of the first four principal components are different. The background spectrum itself contains little information, so only 2 principal components are needed to fully express it. The full spectrum contains more information, so the contribution rate is greater than that of the line spectrum. The first four principal components already cover about 90% of the information in the data.

The analysis of PCA coefficients for the four principal components is shown in Fig. 3.18. The first and second components show opposite trends, with significant coefficients at 64 pixels, 82 ~ 85 pixels, and 105 pixels, corresponding to the line

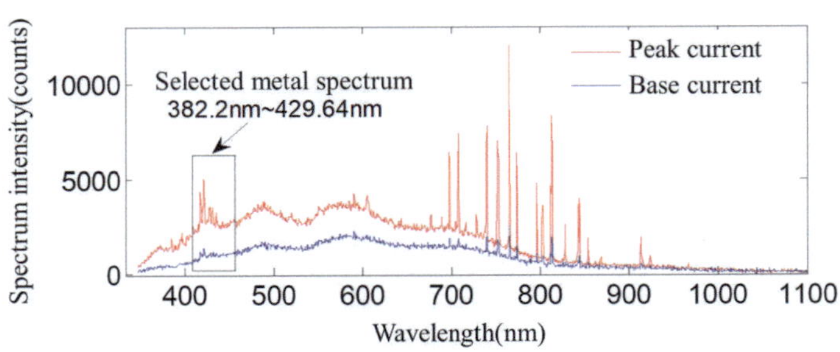

Fig. 3.16 Arc spectrum in different welding current

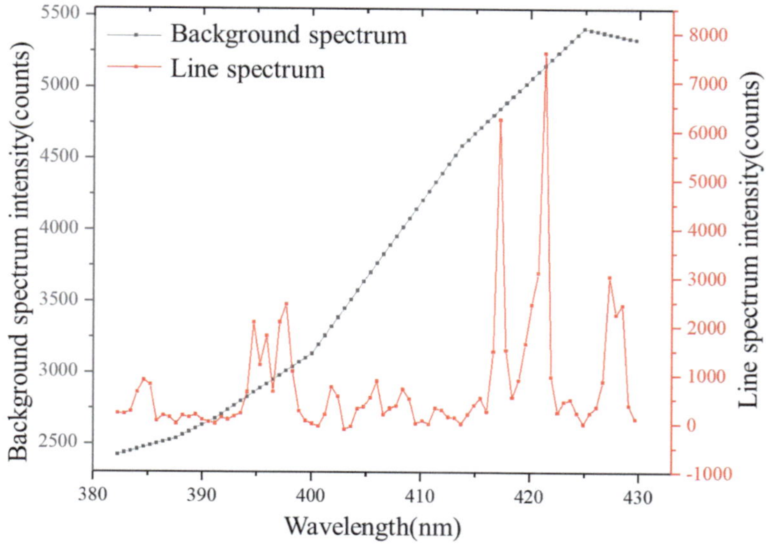

a) Background spectrum and line spectrum

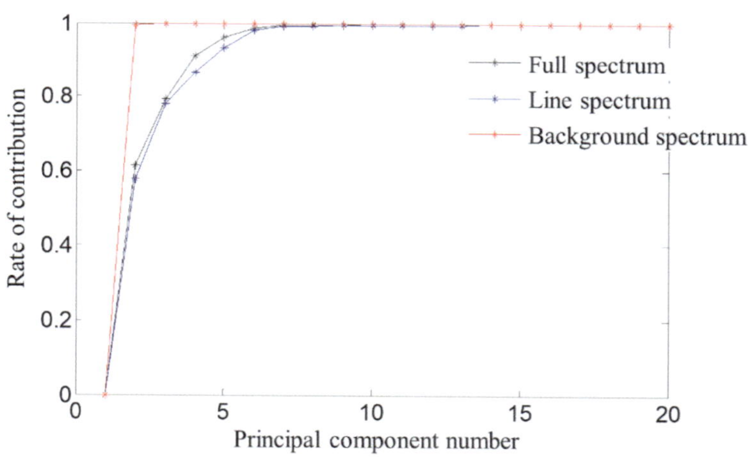

b) PCA contribution rate

Fig. 3.17 PCA processing for metal spectrum

spectra in the original spectral signal, and the coefficients' weights are different. In the coefficients of the third and fourth components, there are many negative correlations. For example, in the third component, 105 pixels are negatively correlated with 64 and 82, with the absolute value of the coefficient being the largest. A similar pattern is also observed in the fourth component. This indicates that the spectral lines corresponding to these three pixels exhibit significant negative correlation, and the four

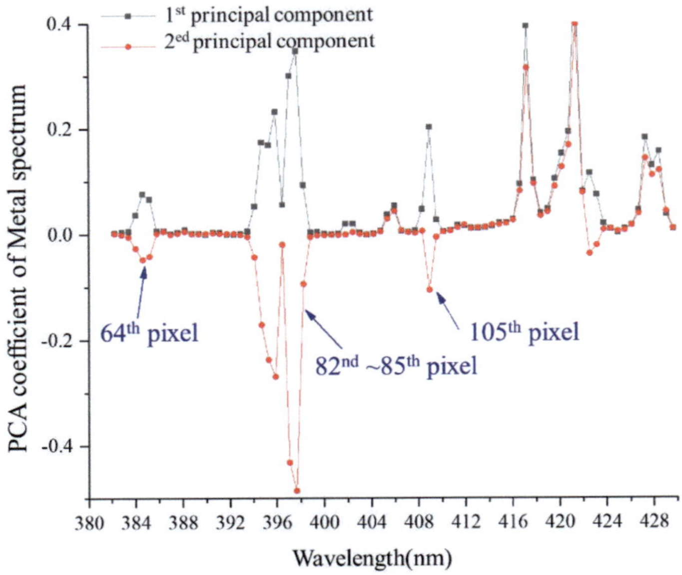

a) Principal component coefficients 1,2

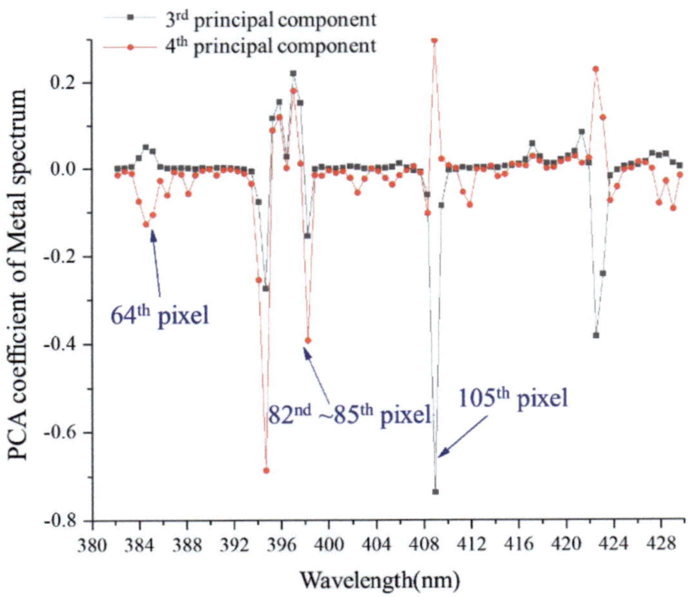

b) Principal component coefficients 3,4

Fig. 3.18 PCA processing for metal spectrum

welding states mentioned above all have different degrees of influence on their radiation characteristics. However, it is still unclear about their unique correlation with process interference factors. This analysis suggests that the first and second components capture complementary information, possibly related to different aspects of the welding process. The negative correlations in the third and fourth components indicate complex relationships between the spectral lines and the welding states, which may be influenced by various factors such as material properties, welding parameters, and environmental conditions. Further research is needed to fully understand these relationships and their implications for process control and monitoring in welding applications.

To further reveal the negative correlation among these spectral lines and clarify their corresponding relationship with the welding process interference factors, Fig. 3.18 compares the PCA main components with the intensity curve of the 105th pixel spectrum. It can be seen that the 3rd principal component (PC) and the 1st PC show opposite trends in the first half of the welding process, corresponding to the no wire feed, wire feed with grinding state of the welding process, indirectly confirming their negative correlation. Furthermore, combining the analysis of the 3rd PC from Fig. 3.19, the coefficient of the 105th pixel is negative and the smallest, indicating that the negative values of the 3rd PC in the second half of Fig. 3.19 are mainly contributed by the 105th pixel, suggesting a significant correlation between the wire feeding state and the 105th pixel. On the other hand, all coefficients in the 1st PC are positive, and the characteristic curve is the result of a linear combination of coefficients of each pixel. Therefore, the metal spectrum demonstrates richness in information and provides a more refined characterization of the welding process compared to the Ar spectrum and its temperature and plasma parameters. This analysis indicates that different spectral lines can provide valuable insights into different aspects of the welding process, such as the presence of contaminants, the stability of the arc, and the quality of the weld. By understanding the relationship between these spectral lines and the welding process, it may be possible to develop more effective monitoring and control strategies to improve welding quality and efficiency.

The spectral line elements with large coefficients were confirmed using the NIST database. Due to the limited pixels of the CCD spectrometer, the confirmation took into account the content of each element in the welding plate and wire, the difficulty of radiation ionization, theoretical strength, and literature reports. The final results are shown in Table 3.8.

Figure 3.20 illustrates the radiation changes of these metal spectral lines in wire feeding and non-wire feeding states. It is evident that the radiation intensity of iron atoms (FeI) significantly increases after wire feeding, while the changes in magnesium (MgI) and aluminum (AlI) are less pronounced. This indicates that the change in wire feeding only affects the FeI spectral line, suggesting a unique strong correlation between the two. The reason for this could be that the welding wire directly enters the arc space and undergoes vaporization and ionization due to the arc's thermal effect. The iron content in the welding wire is approximately 0.38%, equivalent to that in the base material. However, according to the ionization parameters in Table 3.8, its theoretical relative intensity is 5400, which is hundreds of times higher than that of

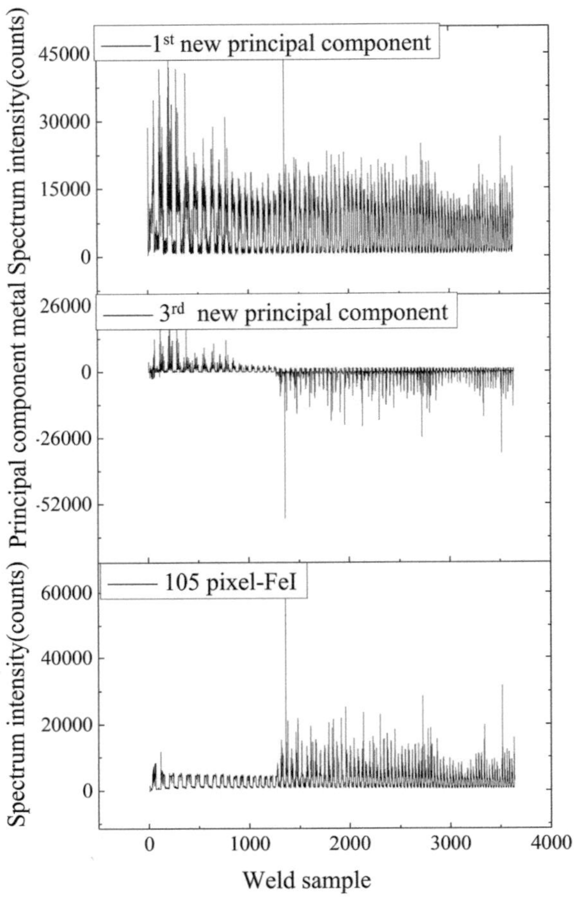

Fig. 3.19 Feature curves of metal spectrum during welding process

Table 3.8 Selected metal lines spectrum using PCA

Spectrum	Wavelength w/nm	Pixel	Relative intensity	Transition probability $A_{ki}^{(s-1)}$	Ionization voltage E_k/cm^{-1}
Al I	396.15	85	26	9.85e+07	25,347.75
Mg I	383.83	64	40	1.61e+08	47,957.045
Fe I	407.84	105	5400	4.55e+06	45,551.76

Al and Mg. This results in the FeI spectral line intensity being much greater than that of AlI and MgI (Fig. 3.19). Additionally, this is also the reason for the sharp increase in the FeI spectral line intensity after wire feeding begins (Fig. 3.20), while the Al element, as the main element in both the welding wire and the base material, does

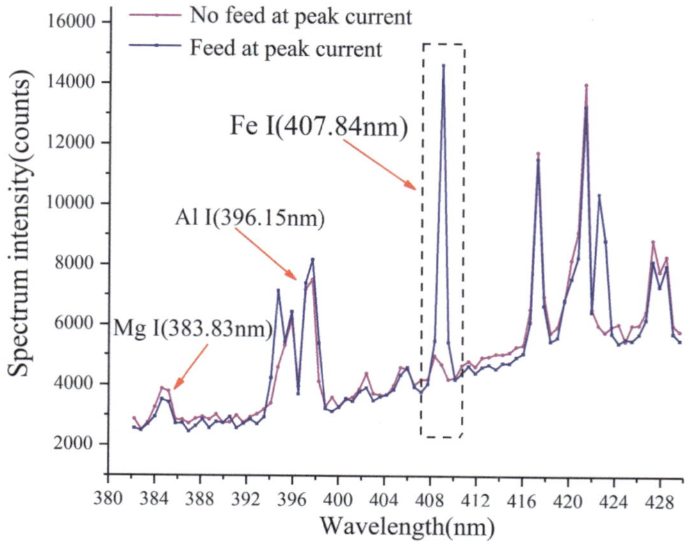

Fig. 3.20 Metal spectrum in different feeding condition

not show significant anomalies. Therefore, PCA has been used to extract the most sensitive and responsive metal elements to changes in process conditions from the high-dimensional metal spectral line space.

3.2.3 *Metal Spectral Line Feature Extraction*

The pulsed gas tungsten arc welding (GTAW) process exhibits significant energy pulse variations. Currently, most studies use signal processing methods to remove pulse interference, but these algorithms are costly and may lack real-time capability. Based on the previous analysis of spectral line dynamics, absolute intensity values are not suitable as process monitoring indicators. The coefficients of the eigenvectors obtained from PCA, as well as the principal components themselves, can be used as feature parameters. However, considering the drift characteristics of the spectrometer itself and the complexity of the welding environment, there is a need to further propose feature extraction methods with higher robustness. Therefore, a method based on the area of interest (AOI) of the metal spectral lines, derived from the PCA coefficients curve, is proposed.

As shown in Fig. 3.21, PCA processing is performed on every 60 spectral samples, and the AOIs for MgI, AlI, and MgI in the first principal component (PC1) of the PCA coefficients curve are selected, and the area under the curve is calculated. This method does not require removal of pulse interference, overcomes the problem of peak pixel drift, improves robustness, and simplifies the feature extraction and signal processing

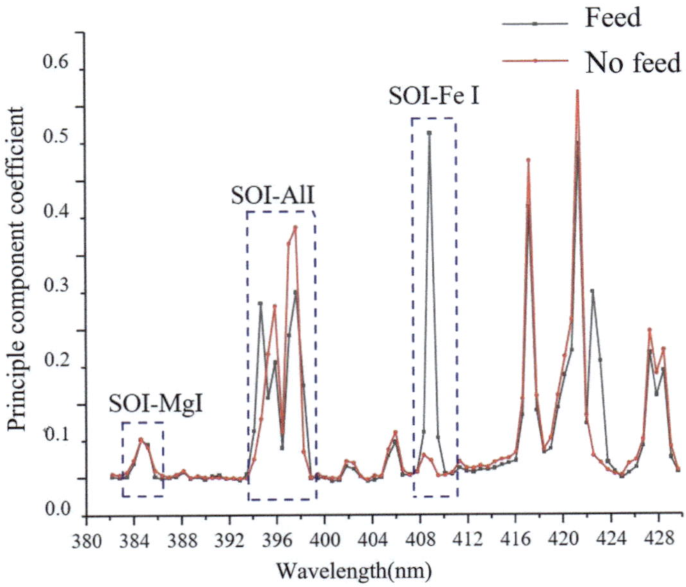

Fig. 3.21 Feature extraction of SOI metal spectrum based on PCA coefficient curves

process. Compared with the linearly combined PCA principal component features, this method can fully demonstrate the positive or negative correlations between metal spectral lines, enabling detailed analysis of the dynamic changes in different metal elements, achieving fine feature extraction, and accurate process characterization.

3.2.4 Metal Spectral Lines Feature Dynamic Responses

Utilization of the proposed process monitoring feature parameters to conduct dynamic response analysis of different welding states, and comparative analysis with the first principal component (PC1) of PCA and original metal spectral line intensities.

The five welding seam states in Fig. 3.22 are as follows: stationary spot welding preheating without wire feeding; walking without wire feeding; wire feeding with appropriate melt width; wire feeding with narrowed melt width; wire feeding with widened melt width.

A. **Analysis of 1–2 Dynamics**

After starting to walk, there is no significant change in the area of FeI coefficient in Fig. 3.23, corresponding to the intensity curve in Fig. 3.24; however, MgI and AlI show varying degrees of increase, with AlI's increase being much greater than MgI's. This indicates that after walking, the ionizing radiation of AlI increased more

Fig. 3.22 Welding seam and its X-ray result

than that of MgI. The reason may be that Al has a higher concentration in the arc atmosphere and a lower ionization voltage than the other two spectral lines, making Al's ionizing radiation more active.

B. **Analysis of 2–3 Dynamics**

After starting wire feeding, the area of the FeI spectral line coefficient increases sharply, reflecting a greater dynamic change in FeI ionizing radiation. There is also a significant peak in the corresponding intensity curve, while the AlI and MgI features show a slight decrease. This indicates that wire feeding has an instantaneous cooling effect on the arc, as the molten wire ionization consumes some of the arc heat. The

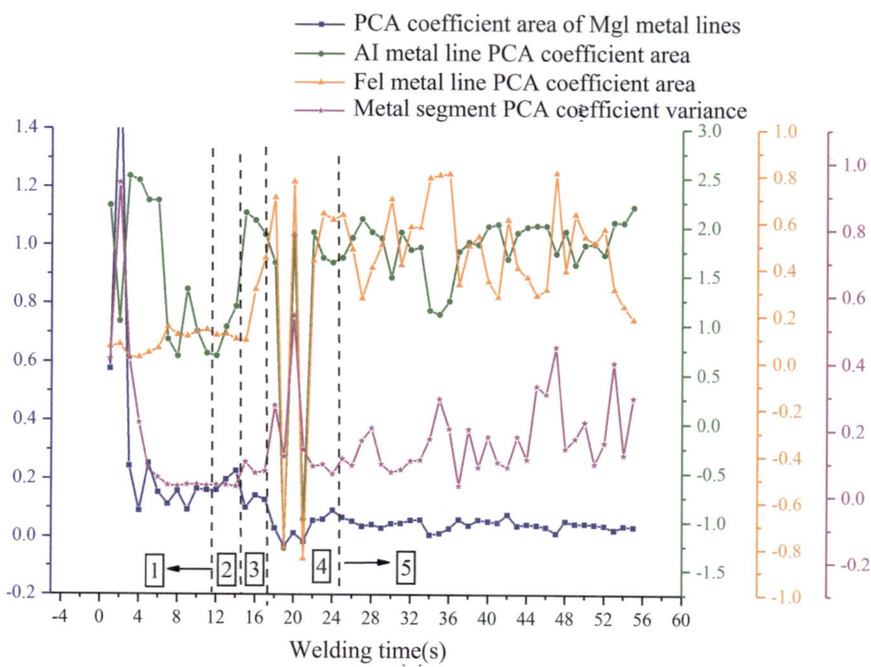

Fig. 3.23 Feature curves for welding process

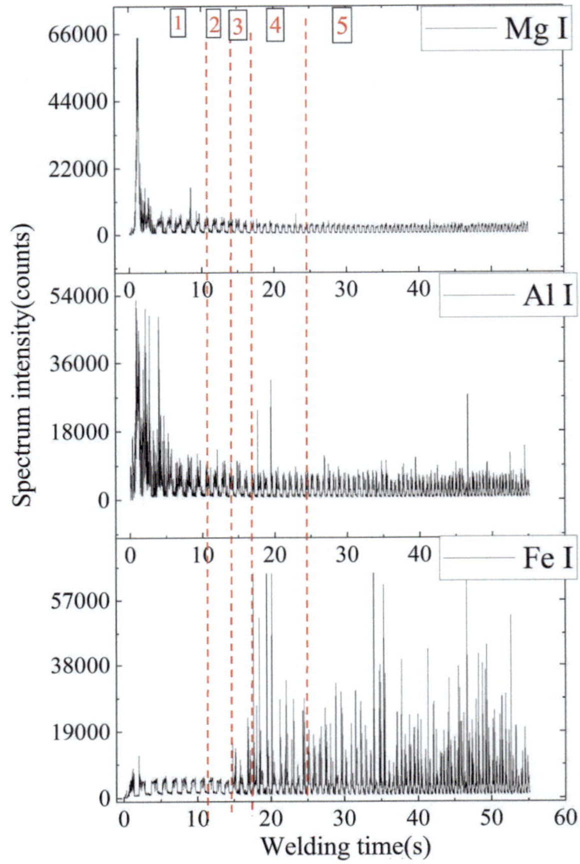

Fig. 3.24 Original spectrum intensity curves

reason may be that FeI has the largest theoretical intensity among the three types of metal spectral lines and the largest PCA coefficient, so the ionizing radiation of FeI increases significantly after the Fe element in the wire enters the arc. Secondly, wire feeding has an instantaneous cooling effect on the arc, as the molten wire ionization consumes some of the arc heat. Additionally, Mg has the highest ionization voltage and a relatively low strength, resulting in the smallest radiation intensity. This validates the negative correlation relationship between the three found by PCA.

C. **Analysis of 3–4 Dynamics**

As the melt depth and width decrease, the effect of heat accumulation decreases, cooling accelerates, and the excitation of the oxide film consumes arc heat. The features of the three types of elements all decrease sharply, with Mg showing the smallest decrease in intensity. Due to its high ionization voltage and the difficulty in ionization, Mg's response is not as sensitive as the other two elements. The increase in the Al component of the wire increases the concentration of Al ionization in the arc atmosphere, increasing its ionization radiation. It can be seen that the dynamic response of Al and Fe is more sensitive than that of Mg. FeI has a relatively large intensity, and the wire is directly exposed to the arc space, leading to full ionization of FeI.

D. **Analysis of 4–5 Dynamics**

As the welding process stabilizes, the MgI feature curve changes little, with its radiation intensity and degree of change remaining constant within a range. AlI and FeI are still active, and their changes are opposite, consistent with the negative correlation phenomenon between the two elements in the PCA coefficient graph. This indicates that different elements have different ionization characteristics and response rules to the same dynamic welding process.

In conclusion, the proposed SOI-PCA coefficient area method has unearthed more potential dynamic change information in the process. Different elements exhibit different dynamic response rules. However, due to the continuity of the process, including transitional states, without clear categories, it is difficult to accurately classify different states.

Figure 3.25 shows the FeI feature parameter curve containing multiple defective welds. It can be seen that under the two states of no wire feeding and welding leakage, the feature curve remains relatively stable, showing no response to welding leakage defects. Only when wire feeding occurs, there is a significant amplitude change, indicating that this FeI (407.84 nm) feature value has a unique correlation with the wire feeding state and is not affected by other factors. The dynamic and rich interaction between the wire and the weld pool makes this element crucial for characterizing the stability and quality of the welding process.

3.2.5 SPC Process Detection

Statistical process control (SPC) method is used for wire feeding status detection in the welding process. The upper control limit (UCL) threshold of the FeI area feature parameter, calculated after normalization, is 4.36. The monitoring parameters in the no-wire-feeding state are all below this threshold. The original, unnormalized 3σ upper and lower limits are 0.137 and 0.0034, respectively. Considering the later fusion with other features for monitoring, the SPC upper control limit method was selected. The test results for multiple welds containing different states are shown in Fig. 3.26. In Fig. 3.26a, only during the wire feeding stage does it exceed the UCL threshold. From the surface of the weld, there are three pre-placed asphalt contaminants, simulating pores and inclusions, but this feature does not respond to simulated defects. In Fig. 3.26b, wire feeding is normal in the early stage, but later the wire gets stuck, resulting in a significant decrease in the feature value with no response to heat accumulation effects. In Fig. 3.26c, no wire feeding occurs throughout the process, and the feature values are distributed near the threshold, showing little response to asphalt interference.

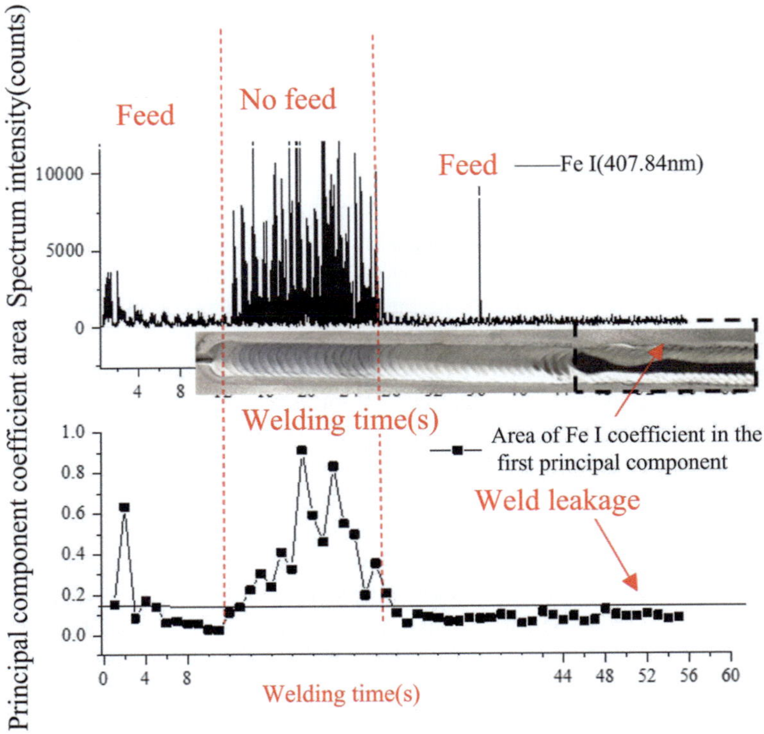

Fig. 3.25 Feature curves for welding seam with multiple defects

3.2.6 Conclusions

In this section, research was conducted on the detection of aluminum alloy pulsed gas tungsten arc welding (GTAW) processes. Through spectral line selection, sensitivity evaluation, feature extraction, dynamic response, and process detection studies on the selected metal spectral bands, the following main conclusions were drawn:

The Fe I (407.84 nm) spectral line has a unique strong correlation with the wire feeding status of the aluminum alloy pulsed GTAW process. FeI (407.84 nm) is negatively correlated with MgI (383.83 nm) and AlI (369.15 nm), showing that wire feeding increases the ionization radiation of FeI in the arc, while reducing the ionization radiation of AlI and MgI.

A feature extraction method based on the principal component analysis (PCA) coefficient of the interested spectral lines was proposed, which has a more refined capability to characterize the welding dynamic process. The dynamic response of Fe and Al is more sensitive than that of Mg.

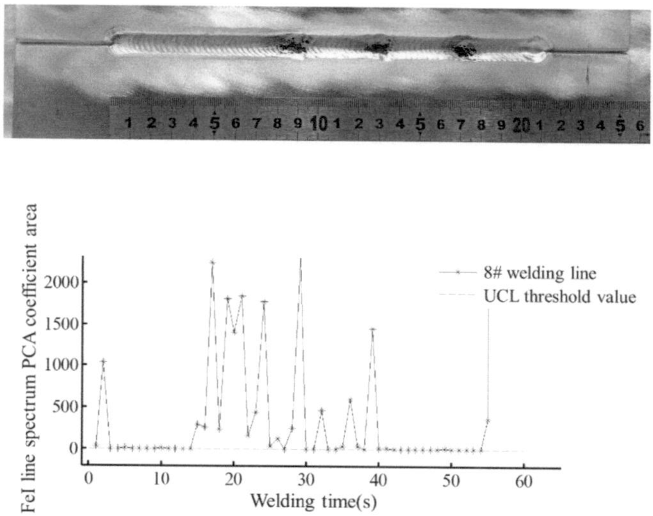

a) Not wire fed in the early stage of welding, and the wire is stuck in the second half

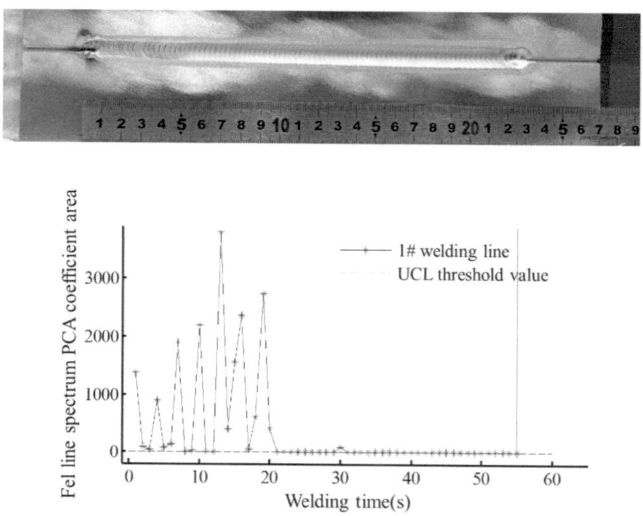

b) Wire feeding in the early stage, and wire clamping after welding

Fig. 3.26 Welding seam and their SPC detection result

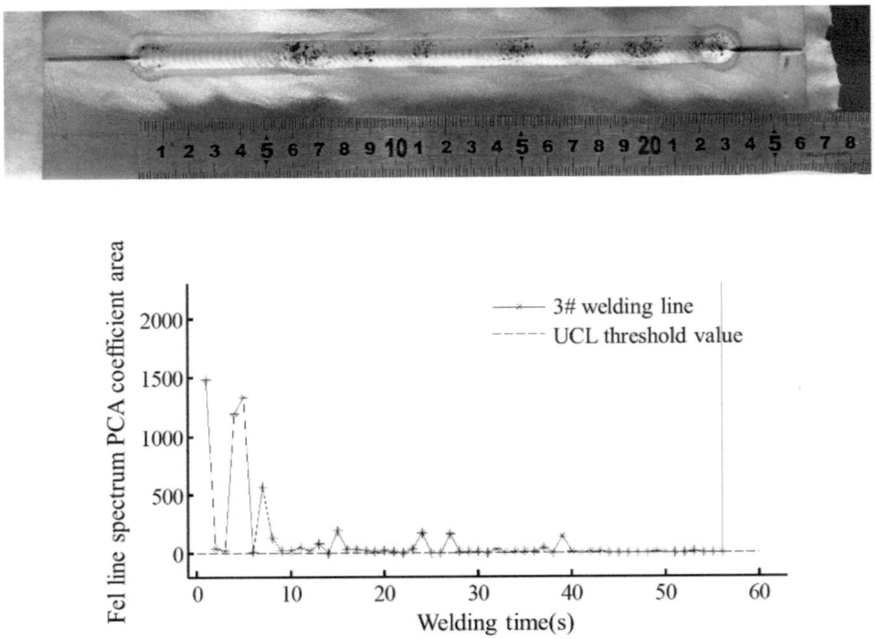

c) No wire fed, preset asphalt

Fig. 3.26 (continued)

Based on statistical process control (SPC), a wire feeding status detection method for aluminum alloy pulsed GTAW processes was established. Through repeated verification in different welding experiments, it was found to have high sensitivity, stability, and strong anti-interference ability.

3.3 Integrating Learning-Based Multiple Seam Defects Detection for GTAW Using Optical Spectroscopy

3.3.1 Experimental Setup and Materials

Figure 3.27 illustrates the experimental setup for robotic welding of Al alloy using Gas Tungsten Arc Welding (GTAW), comprising a robotic welding system, a process-controlling unit, and multiple signal acquisition systems. The Yaskawa industrial robot governed the welding process and speed, while a custom-built in-house monitoring system, operating within the LabVIEW environment, managed arc triggering, signal acquisition, and parameter recording. This monitoring system also regulated wire feeding speed, welding current, and signal sampling frequency. In Fig. 3.27b, an

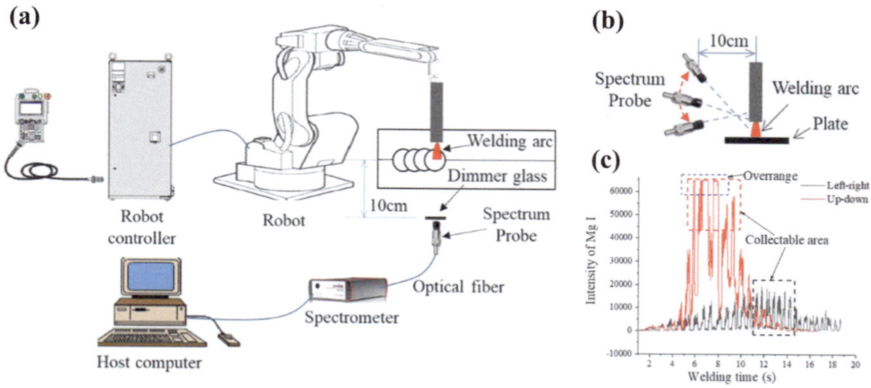

Fig. 3.27 Welding experiment system for robotic GTAW. **a** Robot welding experimental system with optical spectrum acquisition, **b** position adjustment, **c** spectrum intensity curves when moving the probe

AvaSpec-1350F-USB2 optical CCD spectrometer, equipped with an optical probe, fiber, and spectrometer, facilitated optical spectrum signal acquisition. Covering a wavelength range of 350 to 1100 nm with 1350 pixels and a minimum integration time of 0.7 ms, the spectrometer's position significantly impacted the acquired spectrum signal's quality. To address this, the probe was systematically moved horizontally and vertically during welding to determine the optimal acquisition position.

Observations revealed that downward movement of the probe led to partial out-of-range intensity of the Mg I spectrum line (383.83 nm) due to intense arc light. Additionally, slight vertical movement resulted in dramatic spectrum intensity reduction, highlighting the spectrometer's vertical sensitivity, influenced by the welding torch blockage. Consequently, optimal positioning within the collectable area, as delineated in Fig. 3.27c, was essential Conversely, lateral motion exhibited minimal spectrum intensity variation, as the probe faced the welding arc area without obstruction. In the final setup, the probe was fixed 10 cm across the center of the welding arc, with a 10% dimmer preceding it. An integration time of 1.5 ms was set to ensure spectrum intensity around 2/3 of the maximum scale, with a sampling frequency of 60 Hz adhering to the consecutive acquisition model. Once positioned, the probe remained rigidly fastened throughout the welding process to ensure consistent data acquisition.

Table 3.9 enumerates the Al alloy, wire feeding materials, and several pivotal welding parameters. Conversely, Table 3.10 delineates the component weights for the welding material. In the case of 5A06 Al alloy, component weights were determined using an energy-dispersive spectrometer (EDS) for both the base material and the weldment. Notably, the amount of Fe was initially below detectable levels in the 5A06 Al alloy before welding; however, post-welding and SG-AlMg6 wire filling, it measured 0.44. Additional welding experiment parameters are itemized in Table 3.11.

Table 3.9 Welding materials and main parameters

Welding material	Plate size (mm)	Welding wire type/ diameter (mm)	Wire feed speed (cm/ min)	Welding form	Peak current (A)	Welding speed (cm/ min)	Experiment purpose
Al–Mg alloy (5A06)	300 × 50 × 4	SG-AlMg6 Φ1.6	0, 12	Y-grooved butt welding	240	16	Multiple defect detection

Table 3.10 Component weight for the welding plate and wire (wt. %)

	Al	Fe	Mg	Mn	Si	Zn	Ti	Data source
5A06	Balance	–	1.9	–	–	–	–	EDS-tested post welding
5A06 weldment	Balance	0.44	2.68	–	1.07	–	–	
SG-AlMg6	Balance	0.38	5.5–6.5	0.12	0.22	0.05	0.11	Manufacturer data

Table 3.11 Welding experiment parameters

Welding parameters	Value	Welding parameters	Value
Pulse frequency (Hz)	1	Electrode diameter (mm)	3.2
Base current (A)	50	Wire feeding position	Right forward
Pulse duty ratio (%)	50	Feeding angle (°)	45
Ar flow (L/min)	15	Arc length (mm)	4
AC frequency (Hz)	50	Spot welding preheating (s)	8

Figure 3.28 provides an overview of the experimental setups, where spot welding without feeding was conducted post-arc initiation to preheat the plate. Subsequently, post nondestructive testing was conducted using X-rays. For defect simulation, 5A06 Al alloy with Y-grooved butt welding was employed for standard welding. Multiple defects were induced to study online defect detection, including under-penetration, surface and inner porosity, and abnormal wire feeding. Specifically, surface contamination and surface porosity defects were artificially induced by applying preset contamination on the plate surface, resulting in varied degrees of porosity proportional to the contamination level. Inner porosity defects were obtained by neglecting the removal of surface oxidation film before welding, while under-penetration defects were generated by omitting spot welding preheating.

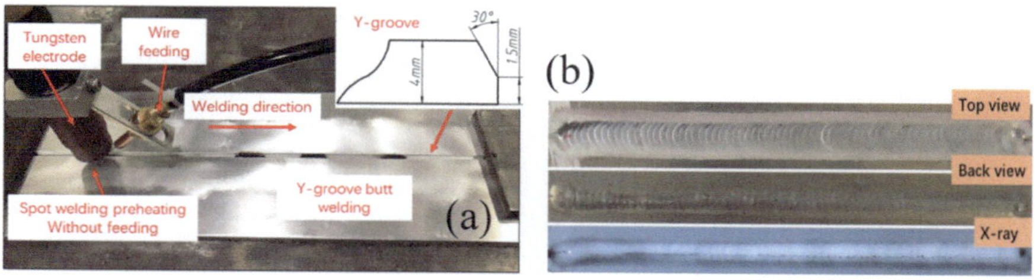

Fig. 3.28 Design of the welding experiments and post welding test. **a** Welding experiments with Y-grooved butt welding, **b** weldment and its test results

3.3.2 Theoretical Background of PCA and Random Forest

Principal Component Analysis (PCA) [1] stands out as one of the most extensively employed algorithms for data dimensionality reduction. In this study, PCA was utilized to assess and select line spectra by analyzing its principal components (PC) vectors and coefficients. The detailed algorithm is elucidated below:

Step 1. The input for PCA should comprise a vector encompassing a specified number of sampled signals recorded under various conditions. In this context, a vector X (N, D) is prepared as the input, where D represents the dimensionality of the spectrum signal, with each dimension corresponding to a specific spectrum wavelength. N signifies the total number of samples for real-time processing.

Step 2. The feature structure decomposition is computed as follows:

$$\lambda_i e_i = Q e_i \tag{3.2}$$

where, $Q = XX^T$ represents the covariance matrix, λ_i represents the eigenvalue associated with the eigenvector e_i.

Step 3. The essence of PCA lies in the linear transformation facilitated by the computed eigenvalues e_i obtained in Step 2. Each dimensional feature, such as the principal component, corresponds to an eigenvalue, and through this transformation, a set of new components with a dimensionality D can be acquired. The linear transformation vector W enables the mapping of the input D-dimensional feature space onto a p-dimensional subspace. Intended for feature reduction, the p dimension of the new feature space should ideally be smaller than D. The resulting new feature vectors Y are characterized as follows:

$$Y_j = W^T x_i; (j : 1; \ldots; P; i : 1, \ldots 1322) \tag{3.3}$$

The eigenvalues e_i also referred to as PC coefficients, are arranged in descending order and constitute the columns of the W vector. Consequently, the new principal components can be derived from the top vectors of Y. Here, p denotes the dimensionality of the new feature space. Typically, p is set to 3, as it is generally observed that the majority of relevant information can be effectively represented by the top 3 principal components.

In this study, PCA and its associated PC coefficients were employed as metrics to quantitatively assess their sensitivity to weld defects and explore the correlation among the one thousand pixels of the spectrum.

Random forest (RF) [2] stands as a cornerstone in integrated learning and machine learning, finding extensive utility across a spectrum of classification and regression tasks. Cerrada et al. [3] innovatively integrated genetic algorithms with random forest to diagnose multiple classes of gear faults. In comparative analyses, RF has consistently outperformed other classification methods, such as logistic model tree (LMT) and classification and regression tree (CART) models, showcasing its efficacy, as highlighted in the report by Chen et al. [4], particularly in map landslide prediction. Furthermore, Manavalan et al. [5] successfully applied RF in pharmacology to predict anti-inflammatory peptides.

At the core of RF lies the decision tree, depicted in Fig. 3.29, possessing several advantageous traits including low computational complexity, swift prediction processes, and intuitive model representation. The RF algorithm can be delineated as follows:

Step 1. Assuming a forest size of k and a training set size of N, each tree is constructed through resampling via the Bootstrap method, independently sampled k times. k sets of training data, each comprising N randomly selected samples, are generated, ensuring that they are IID (independent and identically distributed).

Step 2. Each independent sample set is utilized to construct a classification tree. Given a sample's characteristic dimensionality of M, the growth process of each individual tree entails the random selection of m features from the total M features.

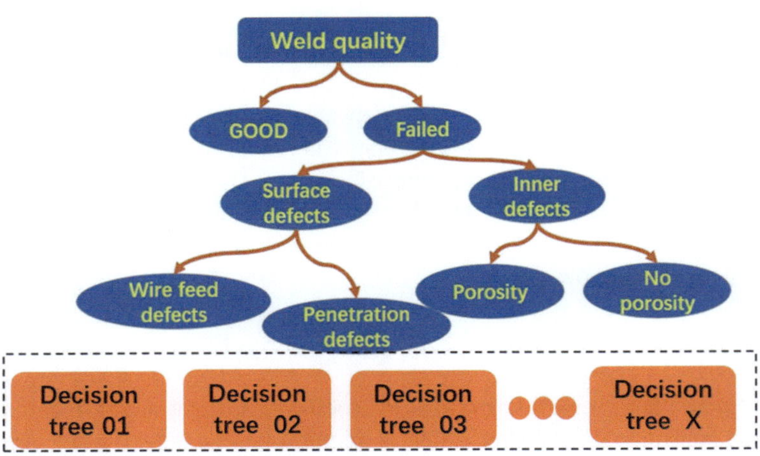

Fig. 3.29 Schematic of random forest for weld defect detection

At each node, the decision tree selects the optimal split from the m features. Each tree is allowed to grow unhindered until the purity of each leaf node is maximized, eschewing any pruning processes.

Step 3. Utilizing the constructed classifier, predictions for new, unidentified samples are made. The classification outcome for each sample is determined through a simple majority voting method applied to the aggregate voting results of all tree classifiers.

In the random forest model, two importance indicators are commonly employed to evaluate the significance of features. One such indicator, known as Mean Decrease Accuracy (MDA), relies on the out-of-bag (OOB) error method. This approach involves randomly perturbing the eigenvalues of the sample data outside the bag and subsequently retesting the OOB error of each tree. MDA is computed as the average difference between the two OOB error tests. Its formula is expressed as follows:

$$J_a(x_j) = \frac{1}{k}\sum_{B_k \in C}\frac{1}{|B_k|}\left(\sum_{i \notin B_k} I\left(h_k^{\overline{x_j}}(i) \neq y_i\right) - I(h_k(i) \neq y_i)\right) \tag{3.4}$$

where: y_i represents the classification label in the i-th OOB, I is an indicator function, and h_k *(i)* denotes the classification label of sample i predicted by the dataset B_k. $h_k^{\overline{x_j}}$(i) stands for the classification label after replacing the feature.

The second importance indicator, termed Mean Decrease Gini (MDG), is founded on the Gini impurity method. It computes the discrepancy between the Gini index before and after classification. The Gini index expression is as follows:

$$Gini(p) = \sum_{i=1}^{n} p_k(1 - p_k) = 1 - \sum_{i=1}^{n} p_k^2 \tag{3.5}$$

where p_k represents the probability of the nth category among the n categories. Its formula is expressed as follows:

$$J_b(x_j) = Gini(D) - \left(\frac{|D_1|}{|D|}Gini(D_1) + \frac{|D_2|}{|D|}Gini(D_2)\right) \tag{3.6}$$

where D_1, D_2 represent the two categories into which the set D is divided based on the features.

It is evident that the indicators in the random forest model can assess the importance of input features. The larger the indicator, the greater the importance of the features. Both methods for evaluating the importance of features are contingent upon the quality of the classification results.

3.3.3 Preprocessing and PCA of Spectrum

Figure 3.30 illustrates signals acquired during the welding process, encompassing pulsed AC-square wave welding current signals and arc spectrum readings at both peak and base currents. This type of pulsed current, commonly encountered in aerospace industry applications, effectively regulates heat input, resulting in superior seam shaping. However, it introduces greater complexity into the monitoring signal, necessitating mitigation efforts.

In Fig. 3.30b, the arc spectrum signal comprises both line and background spectra originating from the liquid weld pool, tungsten electrode, and melted filling wire. To discern between the line and background spectra, lower enveloping is employed, as depicted in Fig. 3.30. Initially, the minimum value is identified to establish the lower envelop of the original spectrum. Subsequently, the line spectrum is isolated by subtracting the lower envelop curve. The separated line spectra, spanning from 360 to 1100 nm, are then utilized as input for PCA for further analysis (Fig. 3.31).

To automatically identify the line spectra closely associated with weld quality and defects, Principal Component Analysis (PCA) was initially conducted on spectrum data obtained from the welding seam of 5A06 during the butt welding process. This dataset encompasses various welding states, such as welding without feeding, with

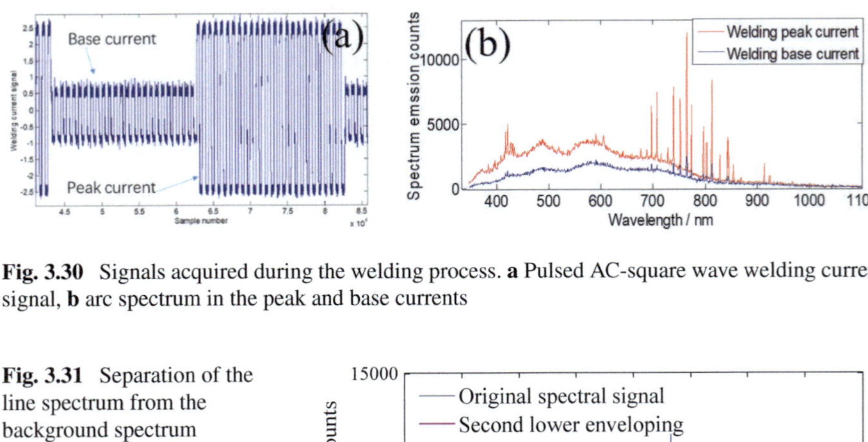

Fig. 3.30 Signals acquired during the welding process. **a** Pulsed AC-square wave welding current signal, **b** arc spectrum in the peak and base currents

Fig. 3.31 Separation of the line spectrum from the background spectrum

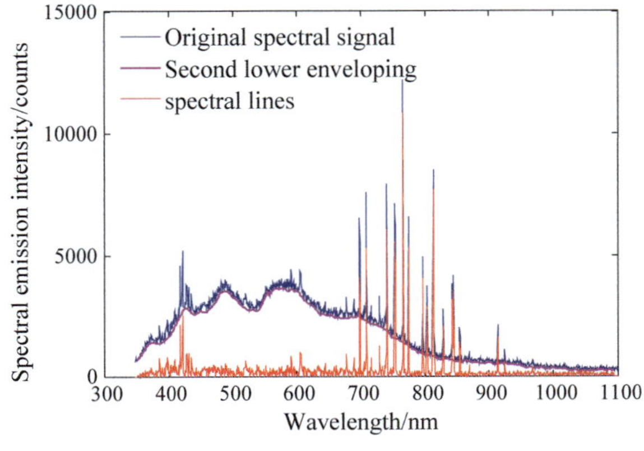

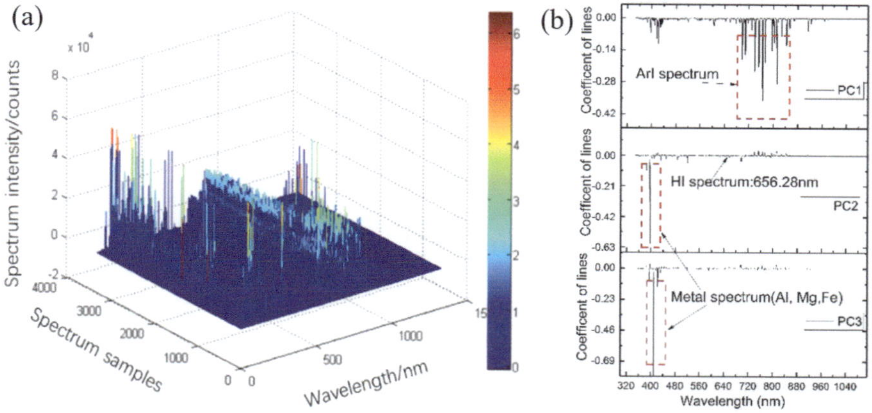

Fig. 3.32 Selected metal line spectrum using PCA. **a** Input of PCA, **b** coefficient of PCs

feeding, porosity, and contamination of the oxidation film. Illustrated in Fig. 3.32a, the total number of samples for the arc spectrum in this welding seam was 3308, with each spectrum possessing a dimension of 1350 corresponding to the spectrum wavelength.

Building upon the principles outlined in Sect. 3.1, the original dimensionality of the spectrum signal, as depicted in Fig. 3.32a, was effectively reduced from 1350 to 3 through Principal Component Analysis (PCA). Initially, the covariance matrix and the PC eigenvectors of the spectrum samples were computed. Subsequently, the three new PC features were derived via a linear combination of each dimension, utilizing the coefficients of the PC eigenvectors. Consequently, the first new PC feature encapsulated the largest variance and contained the most crucial information from the original dataset, followed by subsequent PC features with diminishing variances. These PC features, determined by the PC coefficients, served as quantitative indices for evaluating the sensitivity of line spectra to weld defects, as depicted in Fig. 3.32b.

Upon inspection of Fig. 3.32b, it becomes evident that within PC1, the Ar I line spectrum exhibits a higher value compared to the metal line spectrum present in PC2, encompassing elements such as Al, Mg, and Fe as primary constituents.

The PC coefficients play a pivotal role in the linear combination and feature reduction process. Higher coefficient values indicate greater sensitivity of the line spectrum to weld defects. Hence, it is logical that the coefficient of the Ar I line spectrum, representing the primary PCs of the welding arc, exhibits a higher value. This observation may be attributed to weld defects arising from conditions such as underpenetration or burn-through, which significantly alter the welding arc's length and characteristics.

Conversely, concerning the second PC, it is apparent that metal elements represent other PCs, exerting an opposing effect or cooling influence on the welding arc, as a substantial portion of metal elements absorb arc heat. Although the coefficient of the hydrogen spectrum is relatively low, its association with the porosity of Al alloys has been confirmed [6]. Thus, these line spectra may be closely linked to weld defects stemming from factors such as interference of wire feeding and weld pool contamination.

3.3.4 Feature Extraction for Weld Defects

Figure 3.33 depicts the diagram of the feature extraction algorithm founded upon the selected line spectra, alongside their effectiveness in defect detection leveraging PCA. Specifically, three line spectra with elevated PC coefficients were identified for further analysis: Fe I (407.84 nm), H I (656.28 nm), and Ar I (763.3 nm). These spectra were chosen to investigate their correlation with weld defects, including porosity, abnormal penetration, and abnormal wire feeding.

One of the primary defects encountered in arc welding of Al alloy is porosity, which can significantly compromise the strength of welded products and falls short of stringent standard requirements. The root cause of porosity lies in the substantial difference in the solubility of hydrogen atoms between the liquid and solid weld pools. Given the inherent randomness and complexity of robotic welding processes, porosity cannot be effectively controlled and arises sporadically.

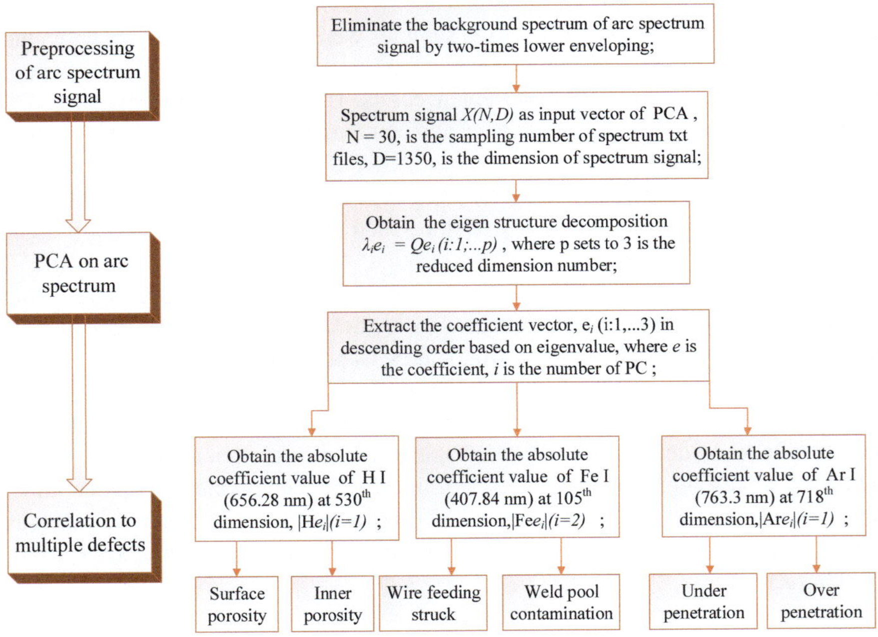

Fig. 3.33 Feature extraction for multiple defects based on PCA

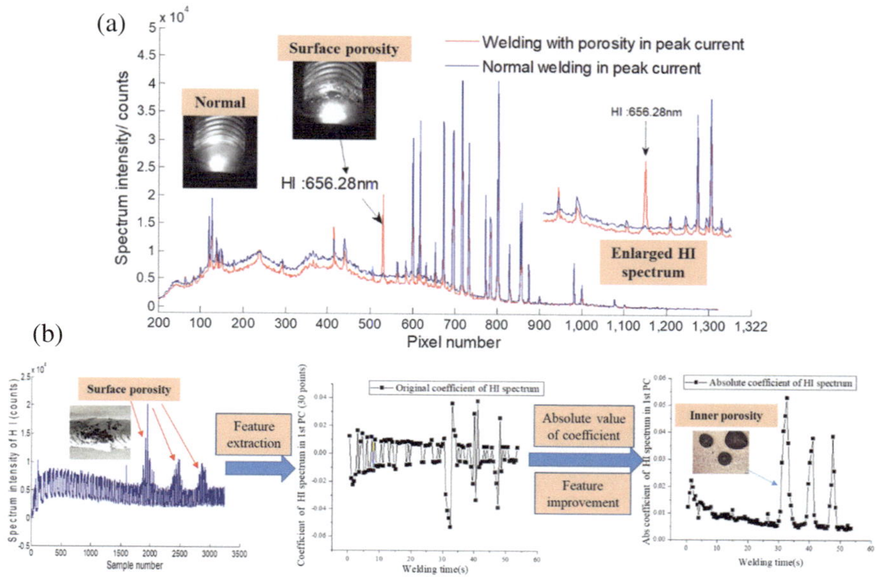

Fig. 3.34 PCA feature improvement. **a** Spectrum signal for weld porosity, **b** absolute PC coefficient of HI (656.28 nm)

In Fig. 3.34a, the acquired spectrum and the H I (656.28 nm) line spectrum are magnified, showcasing a noticeable increase in intensity when surface porosity occurs. Figure 3.34b outlines the feature extraction steps pertinent to seam porosity, focusing on the H I (656.28 nm) line spectrum. Notably, the original intensity of H I exhibits a strong correlation with surface porosity, particularly under significant pulse interference. Subsequently, leveraging the absolute value of the PC coefficient of the H I spectrum remarkably enhances defect detection performance.

In conclusion, the proposed PCA feature, exemplified by the absolute values of PC coefficients, effectively mitigates the noise inherent in PC coefficients and yields superior outcomes compared to the original PC coefficients. Moreover, this method obviates the need for pulse interference elimination and is readily implementable in potential industrial applications.

Figure 3.35 illustrates the correlation of PCA features extracted from Fe I (407.84 nm), H I (656.28 nm), and Ar I (763.3 nm) with the same welding process and weld defects. In Fig. 3.35a, it is evident that surface porosity coincides with over penetration defects, as indicated by spikes in the porosity detection curve alongside the penetration curve. Moreover, the wire feeding detection curve exhibits a distinct response to porosity detection, potentially attributed to preset contamination absorbing some arc heat, thereby reducing the emission intensity of the Fe I line spectrum. Consequently, the detection curve exhibits similar spikes to those observed when wire feeding is stuck.

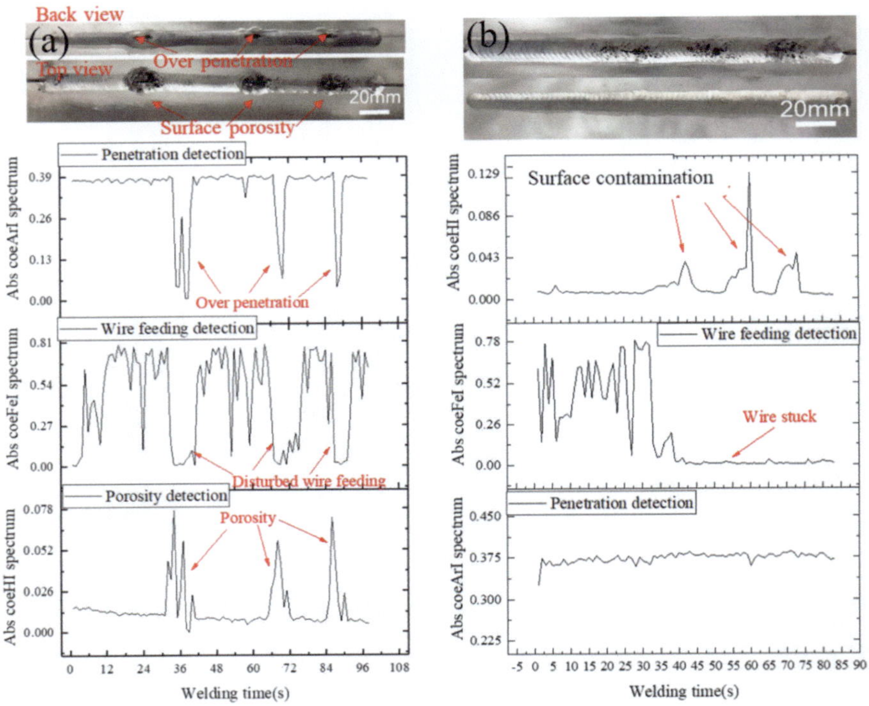

Fig. 3.35 Correlation of PCA features to different weld defects. **a** Surface porosity coupled with over penetration, **b** surface contamination

Figure 3.35b showcases the surface contamination detection results in cases where wire feeding is stuck for a part of the seam. Notably, the detection curves highlight that when defects occur, the width of the Fe I spectrum feature curve is significantly broader than that of H I and Ar I. By computing statistical features, such as kurtosis, their responses to defects were evaluated. The Fe I spectrum feature exhibits the lowest kurtosis value among the three features, suggesting that Fe I emission was impacted for a longer duration by the defects.

3.3.5 Defects Identification Using PCA-RF

The framework of the proposed PCA-RF model is presented in Fig. 3.36, demonstrating a streamlined approach where traditional feature extraction, selection, and model optimization are seamlessly integrated into PCA-RF. Notably, the need for traditional feature extraction in both the time and frequency domains, as well as pulse interference elimination, is eliminated. Instead, only PCA processing on the spectrum is required, proving to be sufficient for subsequent classification tasks.

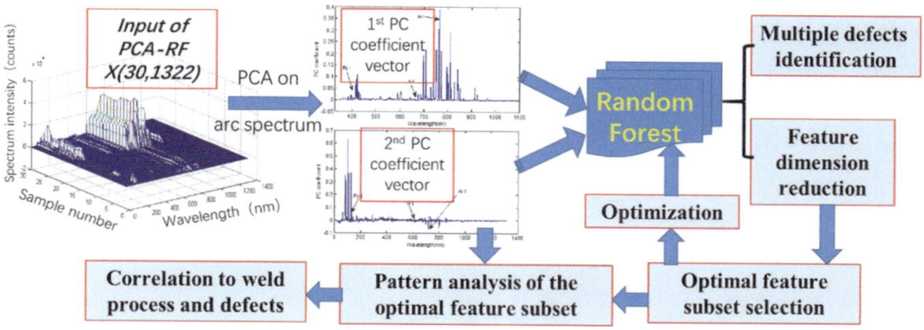

Fig. 3.36 Diagram of the proposed PCA-RF

Moreover, Random Forest incorporates feature selection, evaluation, and classification functionalities within its framework. Consequently, its selection outcomes can qualitatively assess each spectrum element and unveil their heightened correlation to weld processes and defects. This integrated approach not only simplifies the workflow but also enhances the efficiency and effectiveness of defect detection in robotic arc welding manufacturing.

As depicted in Fig. 3.36, the input of PCA-RF comprises 30 samples of spectrum, each possessing 1322 dimensions. Following PCA processing, the coefficient vectors from the first and second principal components (PCs) are both utilized as input features for the RF model. Notably, the spectrum has a sampling rate of 60 Hz, and PCA and feature extraction are executed on every set of 30 samples. Consequently, the proposed method enables inspection of the seam at intervals of approximately 1.3 cm during the welding process.

The output class of PCA-RF consists of six categories: normal seam, coupled defects with surface contamination and overpenetration, wire stuck, surface contamination, inner porosity, and underpenetration. The distribution of samples for training and testing is detailed in Table 3.12. In total, there are 817 samples available for training and testing, comprising 670 samples for training and 147 samples for testing purposes.

To quantitatively assess the importance of each spectrum feature, a novel indicator based on Random Forest (RF) was proposed, namely Important Factor (IF), calculated by determining the variance of the out-of-bag (OOB) test error. As per the theory of RF, Mean Decrease Accuracy (MDA) represents the average difference between two OOB test errors, offering insights into the fluctuation of the OOB test error. Precisely, when a new feature is incorporated into the RF model, minimal variation in the OOB test error indicates that the feature has negligible impact on model performance and is deemed unimportant. Conversely, significant fluctuations in the OOB test error suggest that the new feature holds considerable importance.

The IF serves as a threshold for selecting the optimal feature subset and reducing feature dimensionality. In conclusion, a higher IF value signifies greater importance of the feature in question, thus facilitating efficient feature selection and dimensionality reduction within the RF framework.

Table 3.12 Data description and preparing for PCA-RF

Seam description	Seam appearance	Class (number of samples)
Normal seam		1(402)
Coupled defects	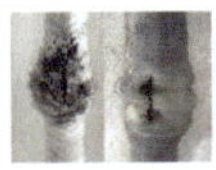	2(55)
Wire stuck		3(113)
Surface contamination	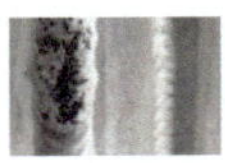	4(51)
Inner porosity	 	5(79)
Under penetration		6(89)

Figure 3.37 illustrates the results of the out-of-bag (OOB) test error during the training and testing phases of the PCA-RF model. The x-axis corresponds to each wavelength of spectral elements, representing the coefficient vector derived from the first and second principal components (PC). Meanwhile, the y-axis denotes the importance of each feature parameter. As depicted in Fig. 3.37, the Important Factor (IF) value ranges from 0.0003 to 0.002, resulting in a reduction of feature dimensionality from 2466 to 12D.

Figure 3.38 presents the tenfold testing results of the PCA-RF model, showcasing the highest accuracy achieved by the PCA-RF defect identification algorithm with an 84D feature set, while the original feature set of 2466D yielded the lowest accuracy. Model performance was evaluated based on standard deviation and mean accuracy, as detailed in Table 3.13. Notably, the model employing the 84D feature set attained the highest accuracy of 0.918 and the lowest standard deviation of 0.018. Interestingly, neither the higher nor the lower feature dimensionality exhibited lower accuracy or higher fluctuation. Higher feature dimensions, such as 213D and 165D, may introduce feature redundancy detrimental to model classification performance. Conversely, lower feature dimensions may yield insufficient information for accurate defect identification.

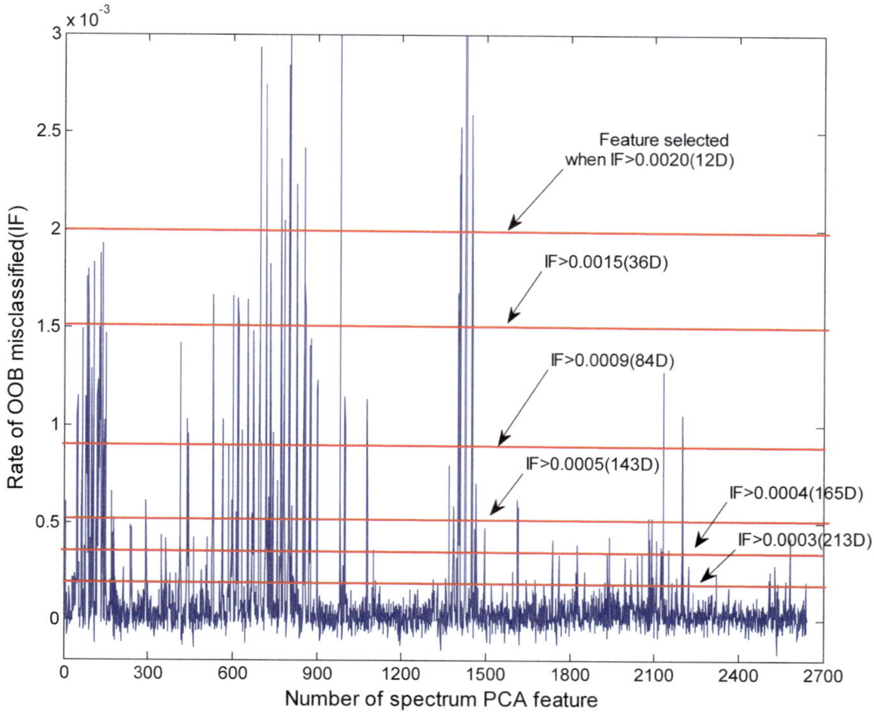

Fig. 3.37 Feature importance and selection based on PCA-RF

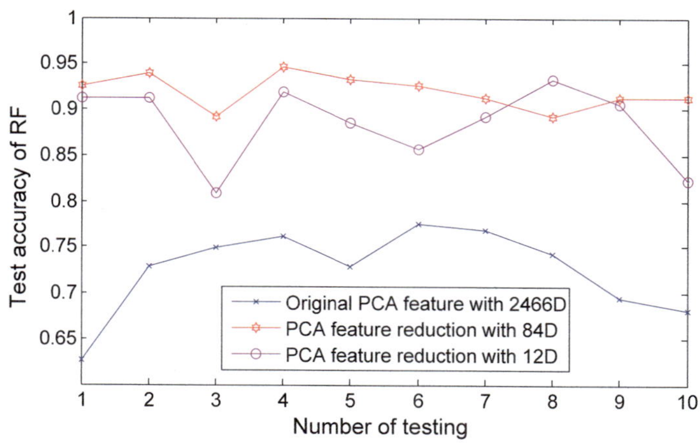

Fig. 3.38 Test curves of PCA-RF with different feature dimension

Table 3.13 Performance of PCA-RF with different feature set dimension

Threshold value of IF	None	0.0003	0.0004	0.0005	0.0009	0.0015	0.0020
Dimension of feature set	2644D	213D	165D	143D	84D	36D	12D
Mean accuracy	0.729	0.908	0.905	0.918	0.918	0.915	0.885
Standard deviation	0.046	0.287	0.030	0.0241	0.018	0.0193	0.039

After the process of feature selection and optimization of the feature subset, the final PCA-RF model was established based on an 84D feature space. Figure 3.38 displays the results of feature importance (Fig. 3.38a), testing accuracy (Fig. 3.38b) with a level of 93.2% accuracy, and decision trees (Fig. 3.38c).

In Fig. 3.39a, while it may be challenging to directly associate each feature with specific spectrum wavelengths and components, more detailed information will be provided and discussed in the subsequent section.

Figure 3.39b reveals misclassifications for defect classes No. 1, 2, and 6. Notably, normal seams were misidentified as coupled defects. Before model optimization, normal seams were frequently misclassified into defect No. 2, No. 4, or No. 6. Compared to the previous classification performance, the proposed model has significantly enhanced accuracy with reduced error rates and standard deviation.

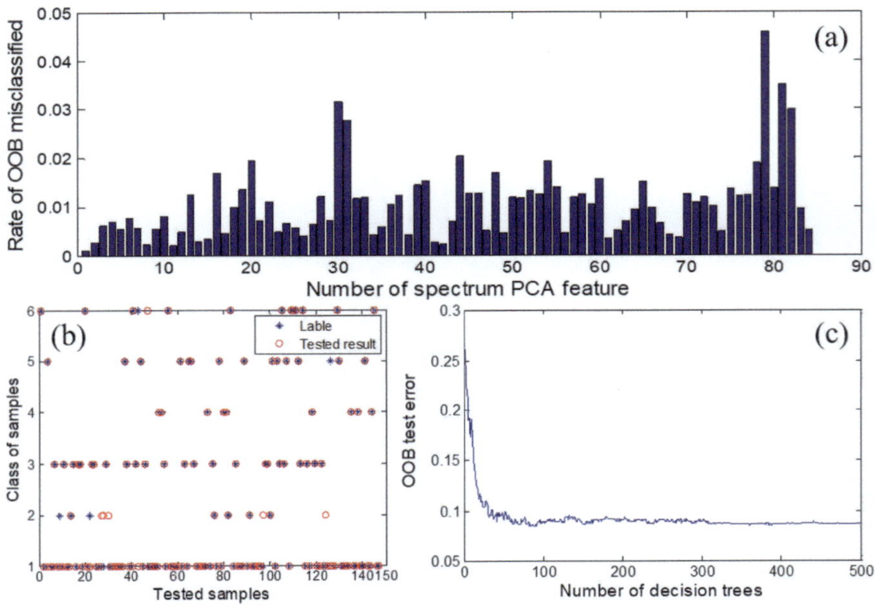

Fig. 3.39 Defect identification based on PCA-RF with 84D feature set dimension. **a** Feature importance, **b** testing results, **c** iteration of decision trees

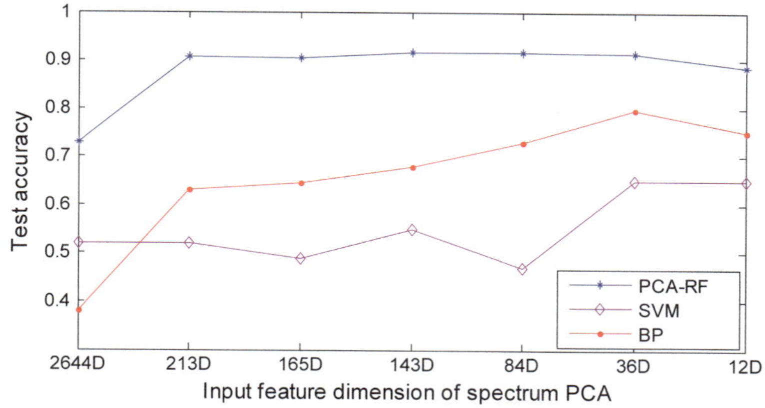

Fig. 3.40 Comparison with different classification models

Figure 3.39c illustrates a decrease in test error as the number of decision trees increases, stabilizing around 300 decision trees. Moreover, PCA-RF exhibits superior defect identification accuracy compared to SVM and BP, while significantly reducing the input feature dimension from 2644 to 12D, as depicted in Fig. 3.40.

3.3.6 Conclusions

In this study, a novel real-time defect identification method, PCA-RF, was introduced for Al alloy in robotic arc welding, leveraging arc optical spectroscopy. Initially, PCA was employed on the arc spectrum to unveil the underlying mechanisms of the dynamic welding process, paving the way for defect feature extraction and detection. It was suggested that spectrum features, particularly the absolute values of the PC coefficients of Fe I (407.84 nm), H I (656.28 nm), and Ar I (763.3 nm), exhibit distinct responses to multiple weld defects.

Subsequently, traditional steps of feature extraction, selection, classification, and model optimization were seamlessly integrated into the PCA-RF algorithm, streamlining the entire process. Additionally, an indicator, the Importance Factor (IF), was proposed, utilizing the OOB test error of RF, to qualitatively assess the significance of each spectrum feature and mitigate the high dimensionality of the feature space.

The PCA-RF method demonstrated notable efficacy in identifying five classes of weld defects or abnormalities, encompassing under penetration, surface contamination, inner porosity, wire stuck, and coupled defects, achieving a mean accuracy of 91.8% with an 84D feature subset. Moreover, it surpassed the performance of SVM and BP methods. The selection pattern of the spectrum feature subset was elucidated and distilled into three discernible rules.

Furthermore, a comprehensive analysis of the selected metal lines spectrum revealed that an increased concentration gradient of Fe, between the wire and the base material, could induce a notable variation in the line spectrum emission of Fe I (407.84 nm).

References

1. A. A. Mohammed, R. Minhas, Q. M. Jonathan Wu, and M. A. Sid-Ahmed, "Human face recognition based on multidimensional PCA and extreme learning machine," *Pattern Recognition,* vol. 44, pp. 2588–2597, 2011/10/01/ 2011.
2. L. Breiman, "Random Forests," *Machine Learning,* vol. 45, pp. 5–32, 2001/10/01 2001.
3. M. Cerrada, G. Zurita, D. Cabrera, R.-V. Sánchez, M. Artés, and C. Li, "Fault diagnosis in spur gears based on genetic algorithm and random forest," *Mechanical Systems and Signal Processing,* vol. 70–71, pp. 87–103, 2016/03/01/ 2016.
4. W. Chen, X. Xie, J. Wang, B. Pradhan, H. Hong, D. T. Bui, *et al.*, "A comparative study of logistic model tree, random forest, and classification and regression tree models for spatial prediction of landslide susceptibility," *CATENA,* vol. 151, pp. 147–160, 2017/04/01/ 2017.
5. B. Manavalan, T. H. Shin, M. O. Kim, and G. Lee, "AIPpred: Sequence-Based Prediction of Anti-inflammatory Peptides Using Random Forest," *Frontiers in Pharmacology,* vol. 9, Mar 27 2018.
6. H. Yu, Y. Xu, J. Song, J. Pu, X. Zhao, and G. Yao, "On-line monitor of hydrogen porosity based on arc spectral information in Al-Mg alloy pulsed gas tungsten arc welding," *Optics and Laser Technology,* vol. 70, pp. 30–38, Jul 2015.

Chapter 4
Optical Spectroscopy-Based Real-Time Seam Strength Prediction for Laser Beam Welding (LBW)

4.1 Monitoring Technologies Used in Intelligent Laser Beam Welding

Laser beam welding (LBW) is a critical process in the aerospace industry due to its high-speed and highly automated nature, which can be affected by random factors, resulting in challenges such as welding instabilities and suboptimal weld seam formation. Real-time monitoring and online detection have become a primary focus of research, leading to the development of intelligent manufacturing. By utilizing extensive data from dynamic laser welding processes [1], it is possible to achieve real-time defect detection.

Real-time data plays a crucial role in controlling laser welding processes, minimizing as-welded defects, and enhancing manufacturing efficiency, precision, and intelligence. Identifying abnormal seams and seam defects in real time allows for evaluating joint strength in each section. While weld seam strength serves as a critical quality indicator alongside macro and internal defects [2], measuring the entire seam strength practically can be challenging, potentially leading to hidden defects that impact component service life and may result in catastrophic failure.

Various sensing technologies have been explored for real-time defect monitoring in robotic laser welding. Feng et al. [3] used digital image correlation to detect in-situ weld bead distortion during tungsten inert gas (TIG) welding of thin plates. A passive vision system captured weld pool images, enabling direct visualization and quality control. Chen [4] used a reflection model leveraged arc light reflection in Al alloy seams to calculate weld pool surface height. Liu et al. [5, 6] investigated vision-based control of weld pool formation using a laser generator and dot matrix structured light. Yang [7] explored real-time detection and location of weld beads based on image analysis and deep convolutional neural networks, valuable for online image-based classification. Additionally, Zhang et al. [8] effectively classified seam penetration of Al alloys during robotic TIG welding using audible arc sound, providing supplementary insights into the welding process.

Z. Zhang et al., *Multiple Sensing-Driven Feature Extraction and Defect Detection*,
https://doi.org/10.1007/978-981-96-4808-5_4

Gu et al. [9] discovered that acoustic emissions ranging from 20 Hz to 20 kHz during CO_2 laser welding carry crucial diagnostic information for weld morphology, penetration depth, and the heat-affected zone. This insight has been applied to analyze various laser welds in mild and galvanized steel. Additionally, Zhang et al. [10] utilized a combination of audible sound, arc voltage, and optical emissions to predict macro weld defects. Cui et al. [11] developed a multi-sensor system to detect welding penetration in keyhole Tungsten Inert Gas welding. In selective laser melting (SLM), Luca et al. [12] crafted a visual sensing method that covers visible and near-infrared emissions to monitor molten pool dynamics, requiring higher acquisition rates and better pixel resolution than TIG welding. While weld pool imaging and arc sound offer limited detection of internal defects or seam strength assessment, Sagar et al. [13] explored the use of fiber Bragg grating sensors for cold crack monitoring in TIG welding, despite the challenges of high temperatures and varying distances.

The arc optical spectrum is recognized as a valuable data source for real-time monitoring and identification of welding flaws. Its advantages include non-contact installation, high sensitivity to welding dynamics, and rich informational content. Pioneering research by Ancona et al. in 2000 [14] and 2001 [15], and further work by Mirapeix et al. [16], used PCA to reduce the spectrum feature space to 14 key wavelengths, have advanced this technology. In 2016, the line-to-continuum ratio of Al I emissions at 396.15 nm became a tool for assessing seam quality in laser welding [17]. Huang et al. [18] explored online monitoring of seam porosity with arc spectrum analysis, while Mazumder et al. [19, 20] examined the composition of metal prints in real-time. Additionally, machine learning has enhanced robotic welding monitoring, with techniques like RF [21], SVM classifiers [8], CNN [22, 23], and extreme learning machines[24] for various quality control aspects.

Accurately predicting seam strength in laser welding remains challenging due to uncertain correlations between the optical spectrum and seam strength, as well as limited training samples. Ensemble learning, an important branch of machine learning, combines simpler models like Random Forest (RF) [21, 25, 26], extreme gradient boosting decision trees (XGBoost) [27–29], and extremely randomized trees (ET) to create more powerful models. These techniques have applications in various fields, including crash frequency analysis [30], antibiotic resistance gene identification [31], ocean color data reconstruction [32], and short-term traffic flow prediction [33]. In 2017, an ensemble data-driven fuzzy network was proposed for quality prediction in laser welding [34], leveraging the Fast Correlation-Based Filter (FCBF) method to extract representative features while minimizing redundancy.

Although some research achievements have been made in recent years, there are still three challenges: (1) The seam strength is influenced by random factors such as matrix or heat-affected zone porosity and post-penetration cracks, making it difficult to accurately predict the seam strength coefficient (SSC) using online monitoring signals; (2) Despite the abundant information contained in the spectrum, which consists of about 3648 pixels and many chemical elements, effective deep mining methods are still lacking to fully utilize this this abundant spectral information. (3) Furthermore, in actual laser welding scenarios, the training samples for different categories of SSC are limited and unbalanced, leading to potential overfitting and difficulties in ensuring robustness.

In summary, achieving accurate evaluation of laser beam welding quality requires obtaining high-precision and high-frequency multi-information, establishing a highly robust evaluation method, to accurately predict the coal seam strength coefficient using online and fully leveraging multi-source information data.

4.2 Extreme Gradient Boosting Decision Tree-Based Seam Strength Prediction for LBW

In this chapter, we investigate a non-destructive measurement method for assessing the strength of laser weld seams. The approach combines principles from spectroscopy and integrated learning. Specifically:

Implementation of Spectroscopic Monitoring System (Sect. 4.2.1):
We detail the practical realization of the spectroscopic monitoring system.

Analysis of Weld Seam Strength Coefficients and Spectral Lines (Sect. 4.2.2):
We explore the relationship between weld seam strength coefficients and spectral features.

Optimal Feature Set Construction (Sect. 4.2.3):
We propose an innovative method that combines random forest and fast independent component analysis to construct an optimal feature set.

Monitoring Results at Different Weld Seam Strength Levels (Sect. 4.2.4):
We present the monitoring outcomes corresponding to various weld seam strength coefficients.

Conclusions (Sect. 4.2.5):
Our findings and implications are summarized in the concluding section.

4.2.1 The Real-Time Monitoring System for LBW

This study presents an integrated system designed for high-precision laser welding of complex weld lines. The system comprises three key components: a versatile robot, an advanced welding system, and a spectrum acquisition module as shown in Fig. 4.1. The experimental parameters is provided in Table 4.1.

1. Robot Platform (Motoman HP20D)

Equipped with six degrees of freedom, the Motoman HP20D provides exceptional flexibility for intricate movements.

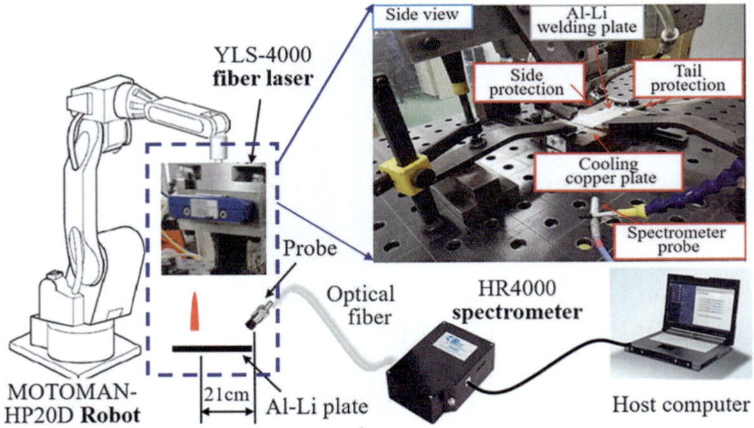

Fig. 4.1 Experiment setup and real-time optical spectrum monitoring system for LBW

Table 4.1 Experimental parameters

No.	Laser power (kW)	Weld speed (m/min)	Plate thickness (mm)	Plate size (mm)
1	4	4	4.5, 3.5, 2.5	400 × 100
2	4	5	4.5, 3.5, 2.5	
3	4	6	4.5, 3.5, 2.5	
4	3.5	2	4.5, 3.5, 2.5	
5	3.5	4	4.5, 3.5, 2.5	
6	3.5	6	4.5, 3.5, 2.5	
7	3	4	3	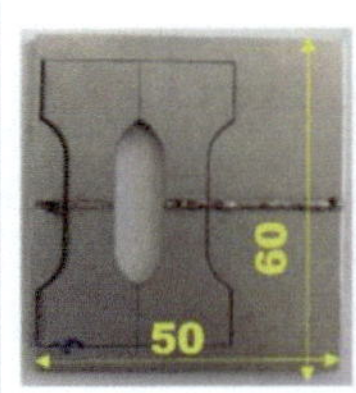 50 × 60
8	2	1	3	
9	2	2	3	
10	2	3	3	

The robot's wrist features an optical fiber laser lens and a precision adjustment holder, enabling precise three-dimensional positioning of the laser head.

2. Laser Source (YLS-4000 Fiber Laser)

The YLS-4000 fiber laser boasts a maximum output power of 4 kW.

With a 2 mm fiber diameter, a spot diameter of 0.2 mm, and a focal length of 146 mm, it ensures optimal beam quality for welding applications.

3. Spectrum Acquisition System

This system captures spectral data during the welding process, allowing real-time monitoring and analysis.

By correlating spectral signatures with weld quality, it enhances process control and defect detection.

In the context of Al-Li alloy welding, effective shielding is essential to safeguard the weld seam from surface oxidation and prevent issues such as cracking, porosity, and softening. This study introduces an integrated spectrum acquisition system designed to enhance the precision and reliability of laser welding processes. As shown in Fig. 4.1, the spectrum acquisition system was comprised of following parts:

1. Shielding Strategy

During Al-Li alloy welding, both side and tail shielding with argon (Ar) gas are imperative. The shielding serves a dual purpose: preserving the weld seam's integrity and mitigating potential defects arising from oxidation.

2. Components of the Spectrum Acquisition System

The system comprises four key elements: a probe, an optical fiber, a spectrometer, and a computer.

The Ocean Optics HR4000 optical fiber spectrometer utilizes a CCD linear image sensor (3648, Toshiba). Its wavelength range spans from 247 to 710 nm, achieving an optical separation rate of 0.1 nm (full width half maximum, FWHM).

Key specifications:

Fiber core diameter: 400 μm.
Spectrometer slit width: 5 μm.
Probe focal distance: 101.6 mm.
Grating lines: 600.

3. Probe Placement and Signal Optimization

The spectrum probe is strategically positioned at the robot's end, aligned parallel to the welding plate.

Maintaining a distance of 21 cm from the laser center, the probe's inclination angle ranges between 30° and 45°.

This configuration balances signal intensity while ensuring minimal interference with robot movement during the laser welding process.

4.2.2 Correlation Between Optical Spectrum and SSC

(1) Theory of plasma emission

Assuming a cylindrical and axisymmetric plasma, with the z-axis aligned along the center line of the plasma column, we define *I(x)* as the radiation intensity of line x from the z-axis. Additionally, let $\varepsilon(r)$ represent the emissivity from central axis of the welding arc r, both $\varepsilon(r)$ and r must satisfy the following relationship:

$$I(x) = \int_{0}^{\sqrt{R^2-x^2}} \varepsilon\left(\sqrt{x^2+y^2}\right)dy \tag{4.1}$$

where R is the distance between the arc edge and the center of z-axis; $r = \sqrt{x^2+y^2}$.

When an electron undergoes a transition from a high-energy state to a low-energy state, energy is emitted at a specific emission rate ε, which can be calculated as follows:

$$\varepsilon(r) = \frac{hcAGn(r)}{4\pi\lambda u(r)}\exp\left(-\frac{E_m}{kT(r)}\right) \tag{4.2}$$

In this context, E_m represents the excitation energy of the higher energy level, *A* denotes the probability of spontaneous emission, which is the likelihood of an electron transitioning from a higher to a lower energy level. *G* is the statistical weight, *n(r)* is the spatial density of the atoms, *u(r)* stands for the partition function, *T(r)* refers to the arc temperature away from the central axis of the welding arc, r being the distance from the axis. The constants *c*, *h*, and *k* represent the speed of light, Planck's constant, and Boltzmann's constant, respectively.

The spectral lines and corresponding chemical elements of the Al-Li alloy are depicted in Fig. 4.2. This figure illustrates a minimal background spectrum when compared to the TIG welding process [10], which is known to exhibit a pronounced continuum spectrum due to black-body radiation emanating from the tungsten electrode, molten feed wire, and the liquid weld pool. In contrast, laser beam welding is characterized by a lower heat input, which is advantageous for welding thin plates while minimizing residual stress and plate distortion. Furthermore, the spectral analysis reveals two distinct types of emission lines: those originating from the base material, including Li I (383.57 nm), Al I (396.15 nm), Mg I (517.27 nm), and Fe I (460.39 nm); and Ar I (610.56 nm), which is attributed to the shielding gas, as confirmed by the National Institute of Standards and Technology (NIST) database.

Moreover, Fig. 4.2 highlights that the Mg I (517.27 nm) spectral line exhibits the most pronounced emission intensity, likely due to its lower ionization energy, reduced boiling point, and greater metallicity. The diverse feature curves of the spectral lines for Li I (383.57 nm), Mg I (517.27 nm), Ar I (610.56 nm), and Al I (670.08 nm), captured during the same LBW process outlined in Fig. 4.3. It is evident that varying

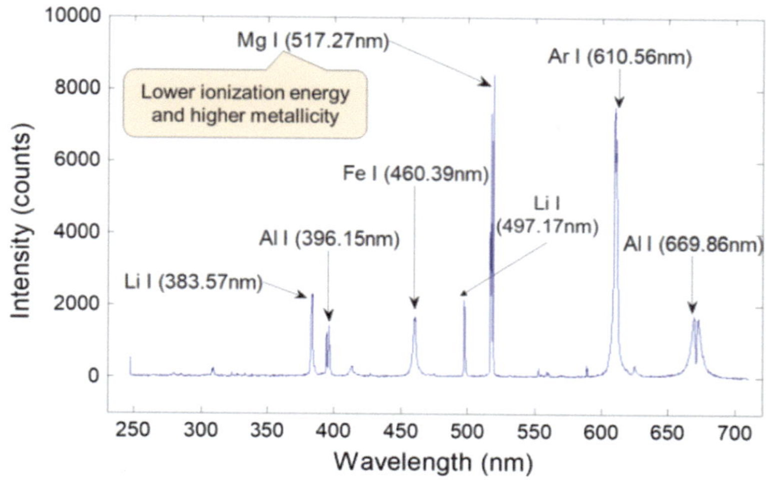

Fig. 4.2 Identified lines spectrum of Al-Li alloy in LBW

emission intensities are displayed by different elements, and these intensities also vary with plate thickness.

To elucidate the relationship between the optical spectrum and SSC, a tensile test was performed on each welded joint. Subsequently, the results from every two samples were averaged to determine the mean SSC value. Figure 4.4 illustrates the correlation between the Mg I (517.27 nm) spectral line and the mean SSC, where the average spectral intensity of Mg I was computed within the designated area of the test specimen. The cubic spline interpolation results indicate a robust linear and positive correlation between the spectral intensity of Mg I and SSC, establishing a vital groundwork for seam strength classification. When Fig. 4.4 is considered alongside Fig. 4.5, it becomes apparent that, in addition to the Mg I element, other

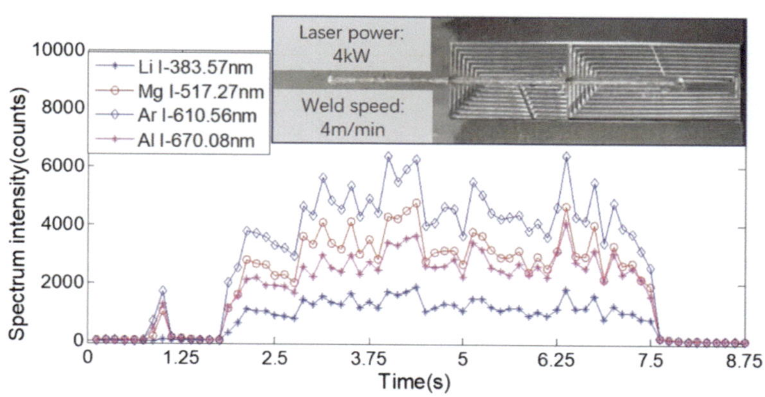

Fig. 4.3 Multiple feature curves of lines spectrum of Al-Li alloy in LBW

elements such as Li I, Al I, Fe I, Mg I, and Ar I also exhibit increased intensity with a reduction in welding speed. This suggests that the laser-induced plasma, resulting from the ionization of the shielding gas and metal vapor, carries significant information pertinent to seam strength.

To elucidate the inherent link between the spectral lines of chemical elements and the strength of the weld seam, an analysis of the original spectral feature set,

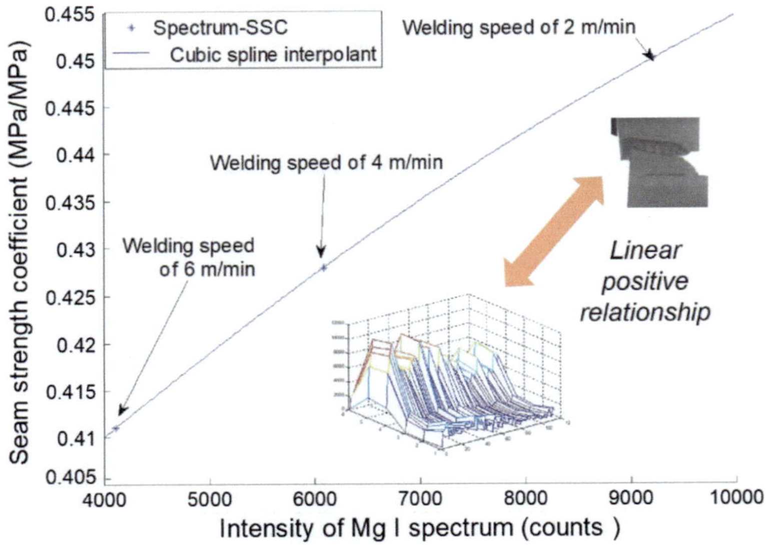

Fig. 4.4 Correlation between the spectrum and SSC at different welding speeds

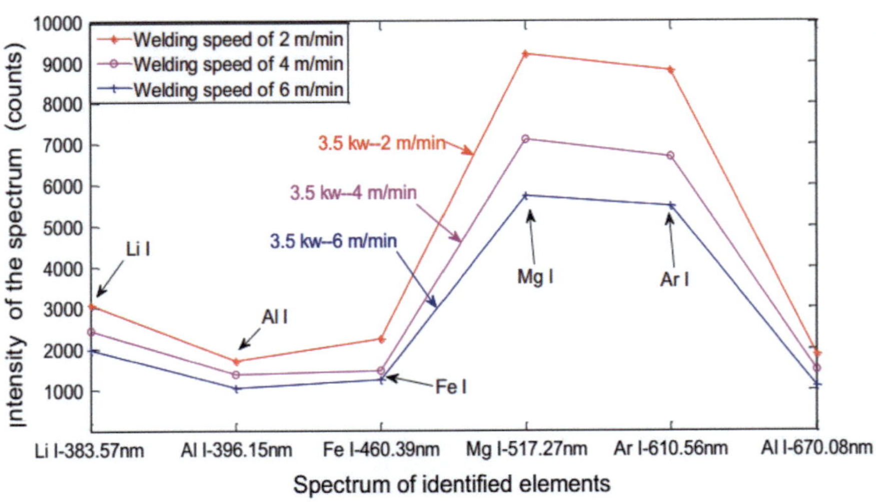

Fig. 4.5 The spectrum of Li I, Al I, Fe I, Mg I and Ar I at different welding speeds

comprising 30 dimensions, was performed. This was based on the feature importance outcomes derived from the Extra Trees (ET) model. The findings are illustrated in Fig. 4.6, where it is particularly noteworthy that the spectral lines of Al I (669.867 nm) and Ar I (610.563 nm) exhibit significantly higher Gini index values compared to other elements such as Mg and Li.

Figure 4.7 presents a schematic representation of the dynamic interplay between the laser-induced plasma and metal vapor during the laser welding process. The plasma generated by the laser consists primarily of two components: Ar atoms ionized from the Ar shielding gas and metallic elements emanating from the base material. It has been verified that the shielding gas imparts a cooling effect on the plasma plume, a phenomenon also captured in high-speed images as depicted in Fig. 4.8. Another contributing factor may be the comparative analysis with other microscale alloy elements.

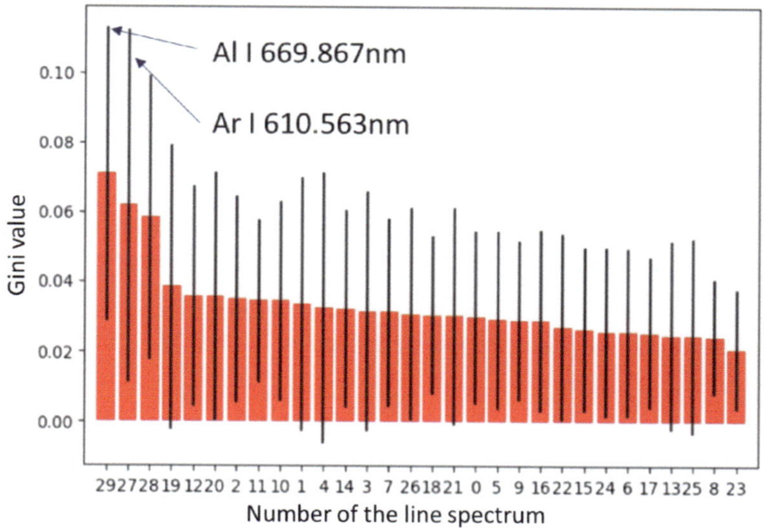

Fig. 4.6 Feature importance of the line spectrum based on ET model

Fig. 4.7 Schematic diagram of the laser induced dynamic plasma

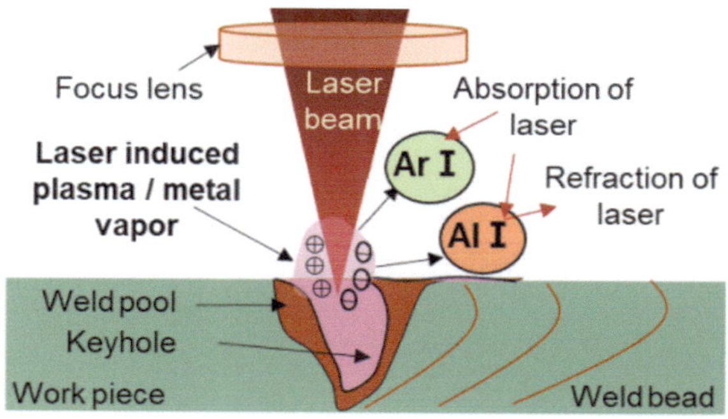

Fig. 4.8 Plasma plume shape observed with high-speed camera

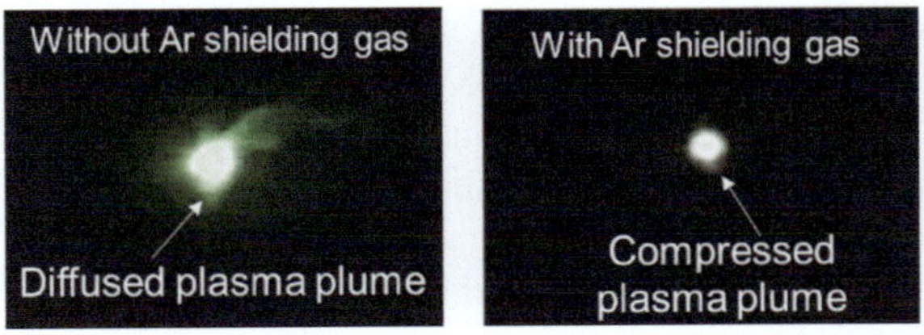

4.2.3 *The RFICA Method to Deeply Mine Optical Spectral Data*

This section delineates the methodology for formulating the optimal feature set through the integration of Random Forest and Fast Independent Component Analysis (RFICA). The process unfolds in five pivotal stages:

(1) Initiate the feature importance evaluation model utilizing RF and MDG, as depicted in Fig. 4.9;
(2) Extract the foremost 30 pixels of the spectrum to establish the new feature vector X, predicated on the MDG values derived from the RF model;
(3) Process the chosen feature set by centering and whitening, resulting in feature vector Wp;
(4) Ascertain the peak negentropy for feature vector Wp via fast ICA;
(5) Convert X into a novel independent feature set S encompassing 10 components.

Figure 4.10 illustrates the workflow of the proposed RFICA-ET. It explicates the essential steps for assembling the feature set and executing SSC classification. The

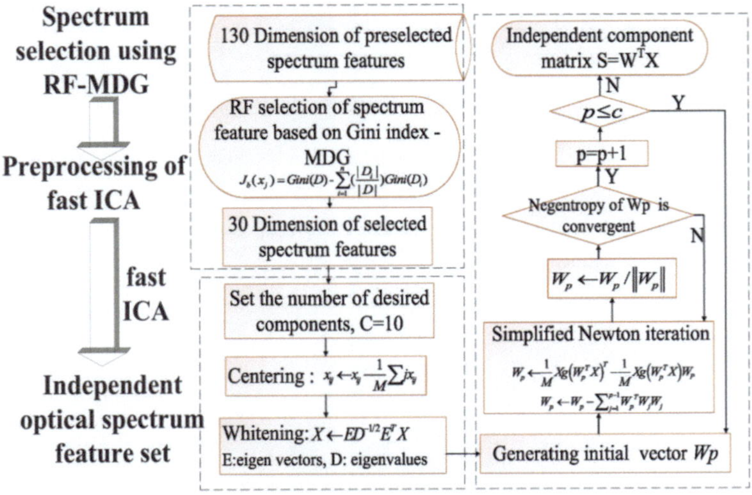

Fig. 4.9 Flowchart of the proposed RFICA

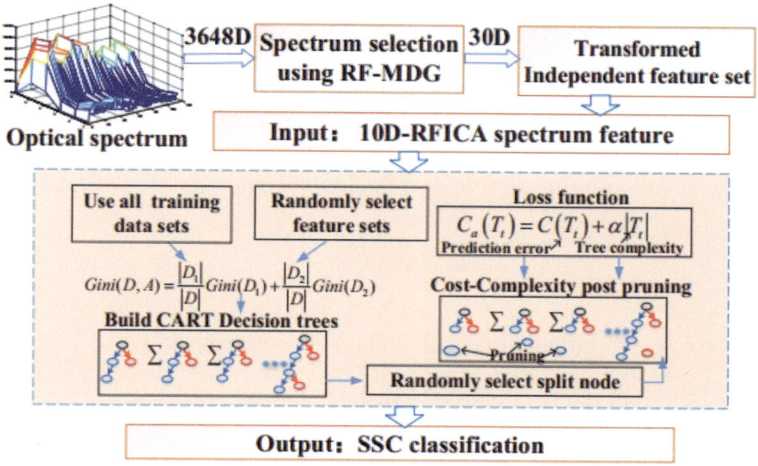

Fig. 4.10 Flowchart of the proposed RFICA-ET

input encompasses optical spectra from N data samples, each with 3648 dimensions, culminating in the seam strength classification outcome.

Step 1: Reduce the feature dimension to 30D and select the optimal feature set based on the RF-MDG value. The feature importance index MDG is calculated as

$$J_b(x_j) = Gini(D) - \sum_{i=1}^{n} \left(\frac{D_i}{D}\right) Gini(D_i) \tag{4.3}$$

Step 2: Transform the 30D feature set into a 10D feature set with high independence and low relevance using fast ICA. Then, evaluate independence of features using maximum negentropy, calculated as

$$W_p \leftarrow \frac{1}{M} Xg\left(W_p^T X\right)^T - \frac{1}{M} Xg\left(W_p^T X\right) W_P \tag{4.4}$$

Step 3. Build CART.

Starting from the root node, the CART is recursively established from the training set, as follows:

(1) Calculate Gini coefficient of sample set D If Gini is less than the threshold value, return sub-tree of the decision tree and stop the current node recursion; Otherwise, proceed to the next step.
(2) Calculate Gini coefficient of each feature of the current node.
(3) Select the optimal feature with the lowest Gini coefficient; Divide the data set into D1 and D2; Establish the left and right nodes of the current node, as follows:

$$Gini(D, A) = \frac{|D_1|}{|D|} Gini(D_1) - \frac{|D_2|}{|D|} Gini(D_2) \tag{4.5}$$

(4) Repeat Step (1)–(3) for the left and right child nodes.
(5) Output CART.

Step 4. Cost-complexity-based post pruning of CART.

(1) Randomly select k features among 10D feature space.
(2) Select a random split and calculate the loss function of each node, as follows:

$$a(t) = \frac{C(t) - C(T_t)}{|T_t| - 1} \quad (4.6)$$

(3) Stop split and return optimal split along with optimal tree T.

Step 5. Determine class of data with majority voting for tree T.

4.2.4 Results Under Different Seam Strength Coefficient

Table 4.1 presents the experimental setup for laser beam welding on Al-Li alloy plates, showcasing two distinct joint types—butt welding and bead welding—and two varying plate sizes. These experiments were designed to validate the proposed methods. The base material's strength reported as 587 MPa by the supplier. The test results pertain to a welded seam width of 7 mm. Subsequently, the seam strength was categorized into four SSC-based classes: 0.2 ~ 0.3 (class 1), 0.3 ~ 0.37 (class 2), 0.38 ~ 0.45 (class 3), and 0.45 ~ 0.5 (class 4). A total of 583 samples were utilized to construct the SSC classification model, with 80% allocated for training and 20% for testing.

Figure 4.11 illustrates a scatter plot comparing two spectral features—extracted via RFPCA and RFICA—from the original line spectrum. Additionally, the Pearson correlation coefficient (Pearson CC) was computed for a quantitative comparison between these features. The Pearson CC registered a minimal value of 0.024 between the first and second independent components of RFICA. This value is significantly lower than those of RFPCA (0.38) and the original feature (0.91), indicating that RFICA can effectively diminish feature dimensions and correlation. More crucially, it can isolate more independent features from the optical spectrum, thereby enhancing feature diversity.

To fine-tune the RFICA-ET model, an exhaustive grid search was conducted to pinpoint the optimal values for two pivotal parameters: the tree estimator and the maximum tree depth (max_depth). It is understood that a deeper tree complexity can lead to more nuanced classification results, potentially boosting accuracy. However, this comes at the expense of computational efficiency. All other parameters were kept at their default settings throughout the optimization process.

The PyCharm 2019.1.3 software was used on a computer running on an Intel Core i7-6560U CPU with 2.20 GHz and 16 GB RAM. Besides, Org-30D refers to the original 30 dimensions of spectrum feature; RFICA-10D refers to the 10 dimensions

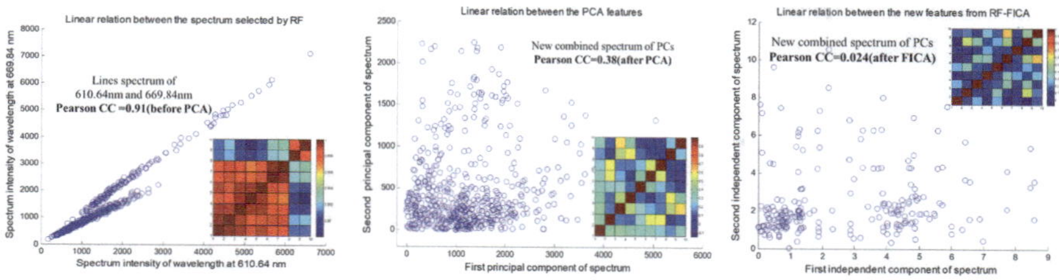

a) The two lines spectrum.

b) The first and second principal components.

c) The first and second independent components.

Fig. 4.11 Scatterplots of the different feature sets and its Pearson correlation coefficient (CC)

using RFICA while PCA-10D referred to PCA; ICPC-20D refers to combing the new features from RFICA and PCA while All-50D were the whole feature sets above.

To compare the effects of tree depth, the tree size was set to 130 and using the RFICA-10D feature set as input. As shown in Fig. 4.12, as the tree depth increases, the accuracy of the RFICA-ET method gradually increases. The highest accuracies were obtained for tree depths ranging from 8 to 20.

The XGBoost model demonstrates the highest stability, particularly at both shallow and deep tree depths, as evidenced by the classification accuracy curves illustrated in Fig. 4.12, with the ET model trailing closely behind. In contrast, the Decision Tree (DT) model yields the lowest accuracy compared to its counterparts.

Informed by the findings presented in Fig. 4.12, we set the tree max_depth to 8 while adjusting the tree size to assess performance variance. As depicted in Fig. 4.13, the novel RFICA-ET approach outperforms the others, showcasing the most robust performance, with XGBoost closely following in terms of stability. With respect

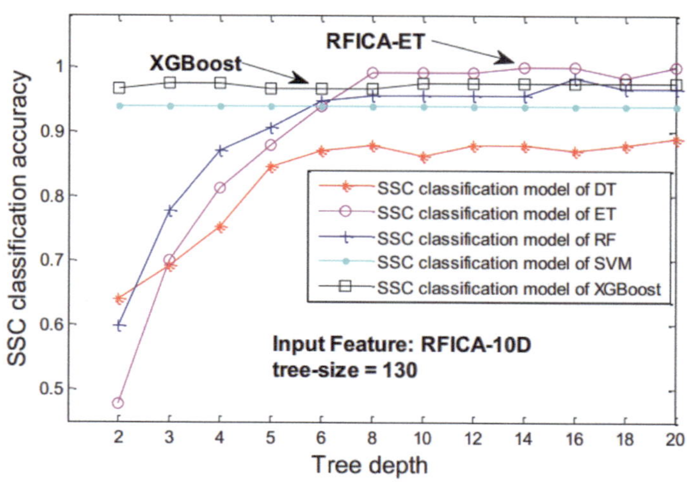

Fig. 4.12 Classification accuracy of different models with varying tree depth

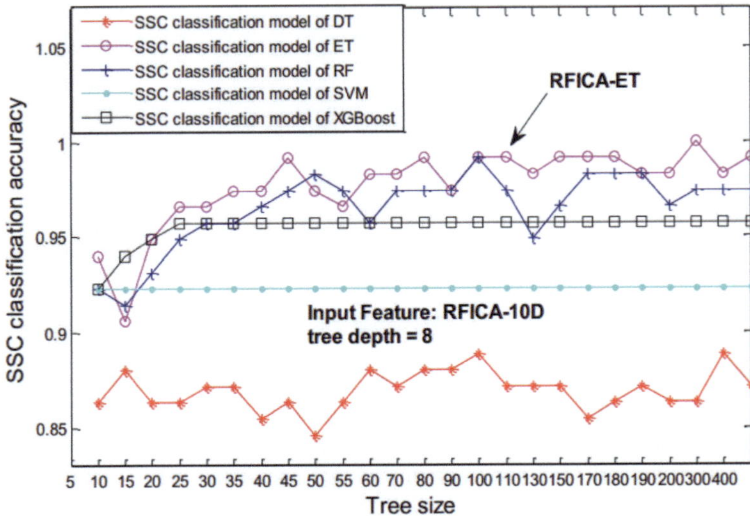

Fig. 4.13 Classification accuracy for different models with varying tree size

to model accuracy, the Random Forest (RF) model ranks second to RFICA-ET. A comparison between Figs. 4.12 and 4.13 reveals that tree depth has a more significant impact on model accuracy than tree size. Consequently, the proposed RFICA-ET classification model is more reliant on the depth of the tree structure rather than its size.

Based on the results of an extensive grid search, optimal values for tree depth and size were determined and applied in subsequent tests. Figure 4.14 delineates the performance outcomes of the RFICA-ET method against DT, RF, SVM, and XGBoost, using a tree depth of 10 and a tree size of 60. It is evident that the RFICA-ET approach attains the highest level of accuracy, followed closely by the RF and XGBoost models. Table 4.2 summarizes the mean classification accuracies of these five models across five distinct feature sets: Org-30D, PCA-10D, RFICA-10D, ICPC-20D, and All-50D. Each model underwent ten iterations of training and testing, with the random_state for data splitting set to non-zero values to prevent any potential bias.

The data presented in Table 4.2 indicates that the ET model achieves the highest mean classification accuracy (0.993) with the RFICA-10D feature set, surpassing the RF, XGBoost, and SVM models, while the DT model exhibits the lowest mean value. The introduction of the RFICA-25D feature set substantially enhanced the performance of all five models. For instance, the mean accuracy for the ET model increased by approximately 64.6% (from 0.6034 with the Org-30D feature set to 0.993). Meanwhile, the mean accuracy for the PCA-10D feature set was 0.8393. Notably, the highest mean accuracy across all feature sets was achieved by XGBoost when utilizing the All-50D features.

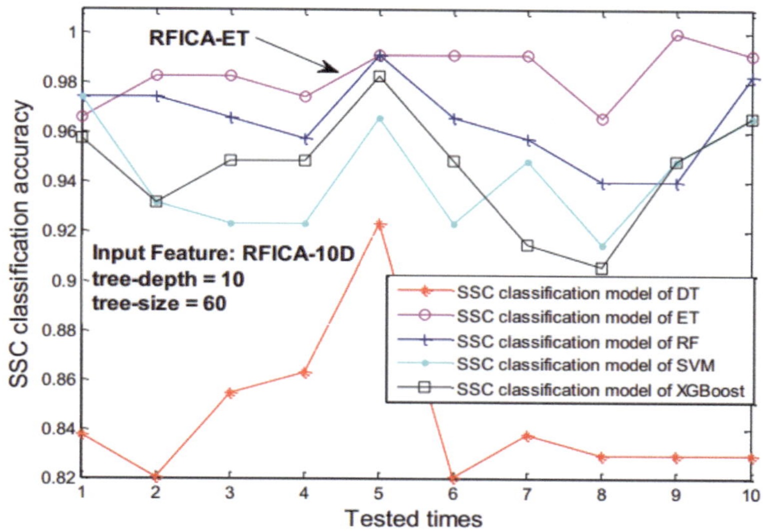

Fig. 4.14 Classification accuracy of different models

Table 4.2 Mean accuracy of different classification models for SSC prediction

Feature set	ET	DT	RF	XGBoost	SVM
Org-30D	0.6034	0.5675	0.5948	0.5863	0.5598
PCA-10D	0.8393	0.5948	0.8	0.7726	0.4940
RFICA-10D	0.993	0.6564	0.9743	0.9453	0.9419
ICPC-20D	0.9786	0.8538	0.9744	0.9427	0.4957
All-50D	0.9427	0.8675	0.9740	0.9521	0.5436

Figure 4.15 shows one of the tree model of the RFICA-ET which can visualize its classification process. The root node, e.g., f8 meaning the eighth feature, is the beginning of pruning for two class. After the simple comparing with the f8, the child node was obtained and moves to judge f6. There is only one tree displayed in Fig. 4.15, however, for the complete model, the predictive algorithm sums the leaf values of all trees to determine the class label.

4.2.5 Conclusions

This study introduces an innovative, nondestructive, and real-time measurement technique for assessing the seam strength of Al-Li alloys during laser beam welding. This method is founded on the use of optical spectroscopy coupled with a sophisticated

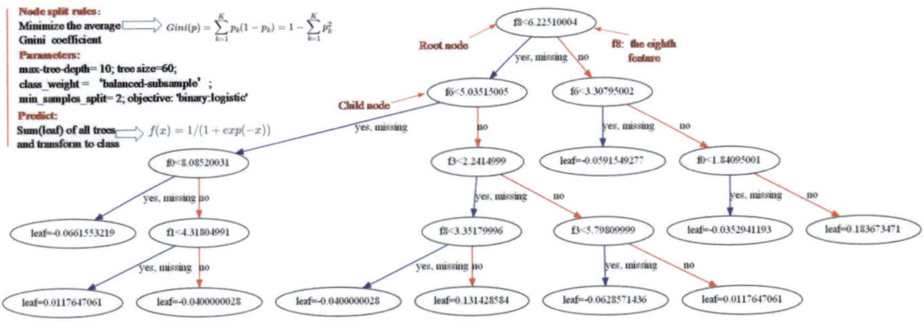

Fig. 4.15 Visualization one of the decision tree of RFICA-ET model

hybrid ensemble learning model. The key findings of the research can be summarized as follows:

(1) The novel RFICA algorithm is capable of deeply interrogating optical spectral data to extract a set of independent features that exhibit lower correlation and enhanced diversity. Leveraging these independent features, the mean SSC classification accuracy of the RFICA-ET model can be significantly improved by approximately 64.6% (from 0.6034 to 0.993).
(2) Among the various tree-based learning models evaluated, the RFICA-ET model (0.993) outperforms others in terms of mean classification accuracy, surpassing RF (0.9743), XGBoost (0.9453), and DT (0.6564). Visualization of one of the decision trees within the RFICA-ET model allows for a clear interpretation of the classification process and its underlying mechanisms.
(3) Driven by spectral data and feature interpretation using the ET model, it was discovered that Al I (669.867 nm) and Ar I (610.563 nm) exhibit notably higher Gini index values compared to other chemical elements, indicating their paramount importance for the accurate prediction of seam strength.

4.3 Ensemble Learning-Based Seam Strength Prediction for LBW

In this chapter, we investigate a non-destructive measurement method for assessing the strength of laser weld seams. The approach combines principles from spectroscopy and integrated learning. Specifically:

Implementation of Spectroscopic Monitoring System (Sect. 4.2.1):
We detail the practical realization of the spectroscopic monitoring system.

Analysis of Weld Seam Strength Coefficients and Spectral Lines (Sect. 4.2.2):
We explore the relationship between weld seam strength coefficients and spectral features.

Optimal Feature Set Construction (Sect. 4.2.3):

We propose an innovative method that combines random forest and fast independent component analysis to construct an optimal feature set.

Monitoring Results at Different Weld Seam Strength Levels (Sect. 4.2.4):

We present the monitoring outcomes corresponding to various weld seam strength coefficients.

Conclusions (Sect. 4.2.5):

Our findings and implications are summarized in the concluding section.

4.3.1 Experiment Design

(1) Laser welding experiment

The detailed parameters of orthogonal experiments are given in Table 4.1, while the plate designed for laser welding to reveal the relationship between the optical spectrum and seam 6 tensile strength as well as for SSC regression modeling is shown in Fig. 4.16.

The study focused on two primary factors: laser power and welding speed. Additionally, the thickness of the Al-Li plates was varied from 2.5 to 4.5 mm in increments of 1.0 mm, along with the seam length for each thickness, as depicted in Fig. 4.16. As outlined in Table 4.1, ten distinct experiments were conducted and repeated between two to five times to ensure a sufficient sample size for training the SSC regression model. The dimensions of the standard specimens utilized in the tensile tests are also detailed in Table 4.1. Prior to welding, a stainless-steel brush was employed to cleanse the base metal surface. This step eradicated the surface oxidation layer and enhanced the base metal's surface roughness, thereby augmenting the absorption of the laser beam. Throughout the welding procedure, the laser beam was angled at 5 degrees to safeguard the optical components. Furthermore, Argon shielding gas was delivered through nozzles positioned at both the front and rear sides of the weld pool, with flow rates of 28 L/min and 30 L/min, respectively, as illustrated in Fig. 4.17b.

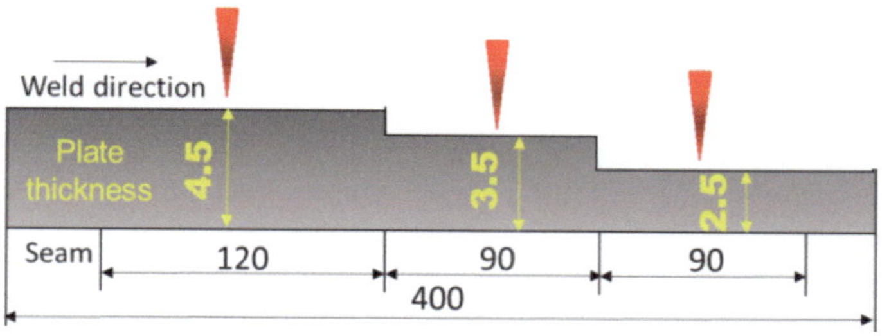

Fig. 4.16 Plate designed for laser welding

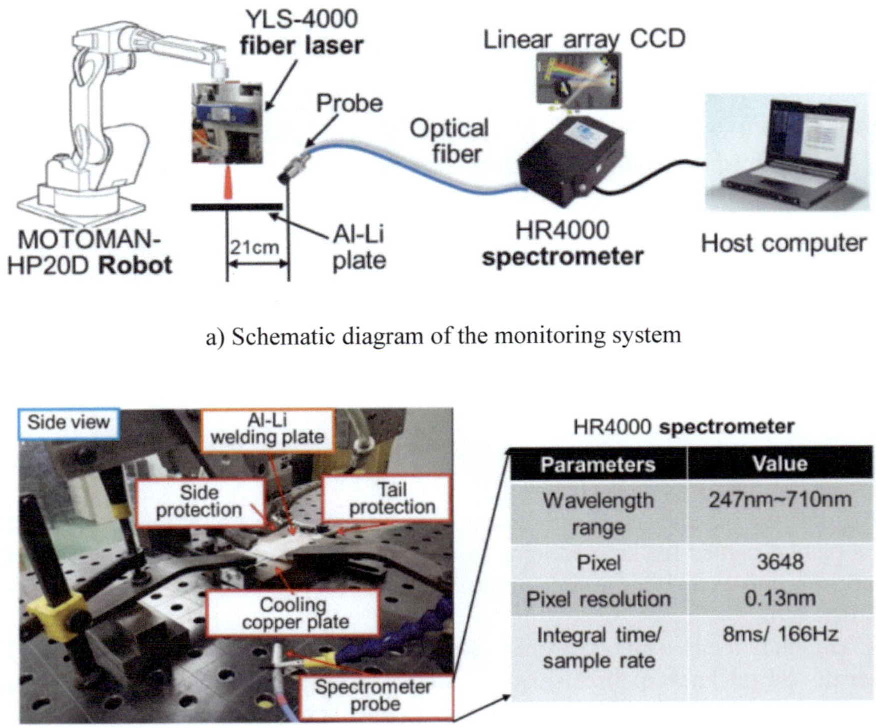

a) Schematic diagram of the monitoring system

b) Side view and spectrometer parameters

Fig. 4.17 Experiment setup and real-time monitoring system of laser welding

As illustrated in Fig. 4.17a, the Motoman-hp20d robot, boasting six degrees of freedom, was employed for the task. The welding operations utilized an IPG YLS-4000 multimode fiber-optic laser system, boasting a peak output of 4 kW and a transmission fiber diameter of 0.2 mm. The system's focal length was set at 150 mm, with a focusing lens diameter of 50 mm, and a focal spot diameter of 0.2 mm. The spectrum acquisition assembly comprised a probe, optical fiber, spectrometer, and a computer for data processing.

As detailed in Fig. 4.17b, the experiment utilized an Ocean Optics HR4000 spectrometer equipped with a Toshiba 3648-pixel linear array CCD, spanning a wavelength range from 247 to 710 nm. Following meticulous calibration, the integration time for spectral capture was established at 8 ms to optimize the signal-to-noise ratio, resulting in a sampling rate of 166 Hz during the laser welding process. The spectral signal exhibited sensitivity to the positioning of the probe. Initially positioned at the robot's extremity, the probe remained relatively stationary relative to the laser beam. Subsequent adjustments involved gradual shifts in both horizontal and vertical planes, culminating in the identification of the optimal acquisition point at a distance of 21 cm from the laser beam's center.

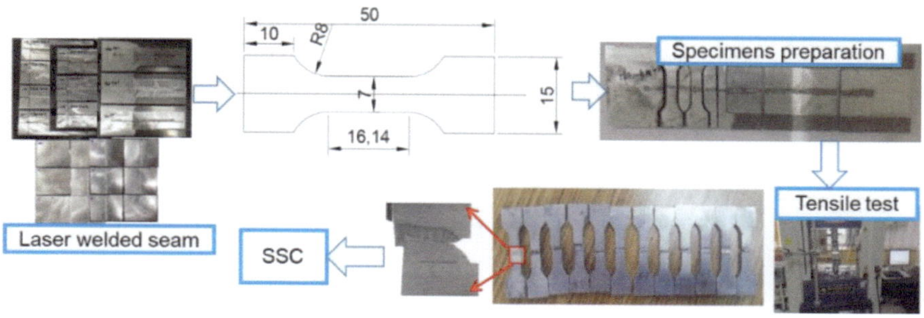

Fig. 4.18 Off-line calculation of the SSC

The detailed steps of the SSC off-line calculation are presented in Fig. 4.18. Two types of welded plates with different values of thickness and size were prepared based on the standard specimen for the tensile test. The SSC was calculated by:

$$SSC = \frac{\sigma_{Tested\ Strength}}{\sigma_{Based\ metal\ Strength}} \tag{4.7}$$

The base metal exhibited a strength of 587 MPa, with each test correlating to a welded seam width of 7 mm. In total, 583 samples were amassed for SSC modeling, with 80% allocated for model training and the remaining 20% designated for testing. Each parameter set underwent a tensile test to yield two strength samples. Across 22 welding experiment groups, 44 tensile tests were conducted in total. It is noteworthy that accurately predicting the SSC for each millimeter of the welded seam proved challenging, especially when attempting to match the sample rates of the optical spectrum. In this study, the strength test outcomes for every pair of specimens were averaged. Furthermore, it was postulated that this average strength corresponded to a seam length of 1.4 cm, along with the associated spectral signals.

4.3.2 *Optical Spectrum Feature Extraction and Optimization*

To achieve online regression and prediction of the Solid Solution Content (SSC) for the Al-Li alloy laser welding process, the preliminary work is divided into two parts: (1) determining the relationship between the optical spectrum and SSC; (2) feature extraction and selection of the optical spectrum signal using the proposed RFPCA method.

(1) Correlation of optical spectrum and laser welding SSC

This section examines the impact of key laser welding parameters on the spectral signal prior to establishing the correlation between SSC and the spectral signal.

The optical spectrum data collected during the Al-Li alloy laser welding process is depicted in Fig. 4.19. Due to the high concentration of the laser, the spectral signal exhibited minimal background spectrum in comparison to other welding methods, such as TIG, resulting in more pronounced spectral peaks. Assuming that the plasma is cylindrically axisymmetric with the arc column's centerline along the z-axis, the radiation intensity at a distance *x* from the z-axis is denoted as *I(x)*, and the emissivity at a radial distance r from the arc center is $\varepsilon(r)$, the following relationship holds true:

$$I(x) = \int_{0}^{\sqrt{R^2-x^2}} \varepsilon(\sqrt{x^2+y^2})dy \tag{4.8}$$

In the given context, R represents the distance between the edge of the arc and the center of the z-axis, and the relationship between the coordinates is defined by the equation $\sqrt{x^2+y^2} = r$.

When an electron transitions from a higher energy level to a lower one, it emits energy at a specific emission rate, denoted by ε, which can be calculated using the following formula:

$$\varepsilon(r) = \frac{hcAGn(r)}{4\pi\lambda u(r)} \exp\left(-\frac{E_m}{kT(r)}\right) \tag{4.9}$$

In this context, E_m represents the excitation energy of the upper energy level, *A* denotes the probability of spontaneous electron transition at that level, *G* signifies the statistical weight, *n(r)* indicates the spatial density of the atom, *u(r)* stands for

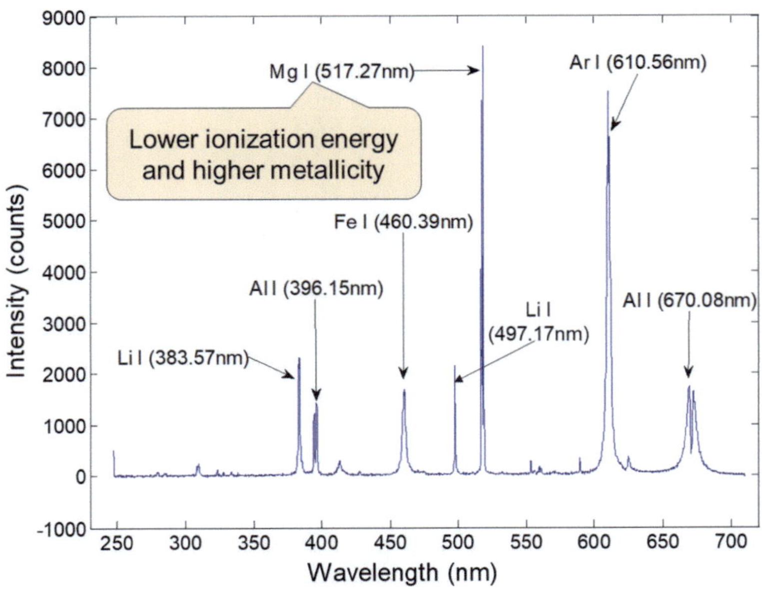

Fig. 4.19 The optical spectrum data of the laser welding process of Al-Li alloy

the partition function, and $T(r)$ denotes the arc temperature at a distance r from the center of the arc. Additionally, c is the speed of light, h is Planck's constant, and k is Boltzmann's constant.

As illustrated in Fig. 4.19, line spectra for various chemical elements were identified, including Li I (383.57 nm), Al I (396.15 nm), Mg I (517.27 nm), and Fe I (460.39 nm) from the base material, as well as Ar I (610.56 nm) from the shielding gas, based on the National Institute of Standards and Technology (NIST) database. Notably, the Mg I line spectrum (517.27 nm) exhibited the highest emission intensity, likely due to its lower ionization energy, lower boiling point, and greater metallicity compared to other metallic elements. The ionization patterns of the shielding gas and metal vapor provide insights into the seam strength.

The observed highest emission intensity of Mg I (517.27 nm), corresponding to pixel number 2069 in Fig. 4.19, prompted an investigation into its relationship with key welding parameters. As depicted in Fig. 4.20a, altering the input laser power while maintaining a constant welding speed of 4 m/min revealed that the Mg I line spectrum reached peak intensity at a welding power of 4 kW across all three material thicknesses (4.5, 3.5, and 2.5 mm). However, the Mg I intensity at powers of 3 and 3.5 kW exhibited negligible differences. Additionally, as demonstrated in Fig. 4.20b, variations in plate thickness significantly affected the spectral intensity. The calculated mean standard deviation (STD) indicated that greater thickness resulted in higher intensity and increased fluctuation. Consequently, thinner plates led to reduced optical spectrum emissions and diminished intensity fluctuations. These outcomes may be attributed to variations in metal vapor concentration during the laser welding process. Specifically, at identical laser power and welding speed, a thicker plate likely generates more metal vapor, which in turn leads to enhanced ionization and optical emissions.

As illustrated in Fig. 4.21, in addition to the Mg I element, Ar I (610.56 nm) also exhibited a higher degree of ionization and radiation than the other four elements. The results were similar to those presented in Fig. 4.20; that is, as the laser power increased or the welding speed decreased, the radiation intensity of the six elements gradually increased. However, the influence of laser welding power on spectral intensity was found to be greater than that of welding speed, as evidenced by the results in Fig. 4.21a and b.

The mean Solid Solution Content (SSC) was calculated for three repeated groups of the welded seam, with each seam represented by two test specimens. As shown in Fig. 4.21a, when the laser power was increased from 3 kW or 3.5 kW to 4 kW, the mean SSC demonstrated a negative correlation with the spectrum intensity. Conversely, a positive correlation between the mean SSC and spectrum intensity was observed when the welding speed was increased, as indicated in Fig. 4.21b. This inverse relationship highlights its dependency on the welding parameters. While the optical spectrum typically has a linear positive relationship with laser power, seam strength is also influenced by several other factors and has its own parameter window. If the parameters exceed the normal range, such as in Fig. 4.21a, an increase in spectrum intensity while the SSC decreases is expected, as the SSC is strongly related to the set parameters.

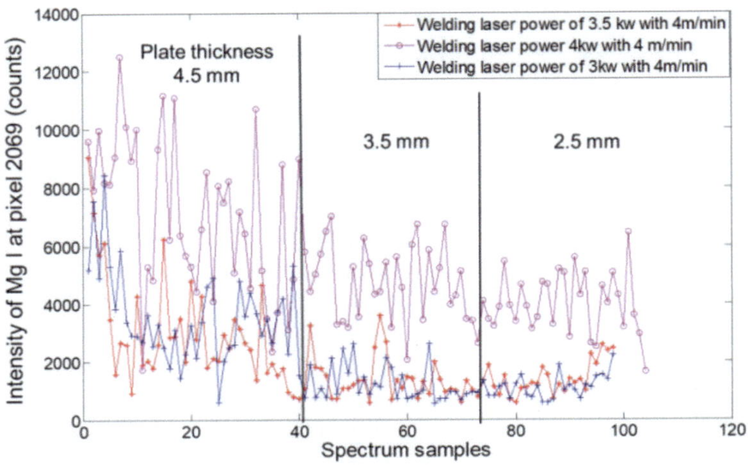

a) Welding laser power

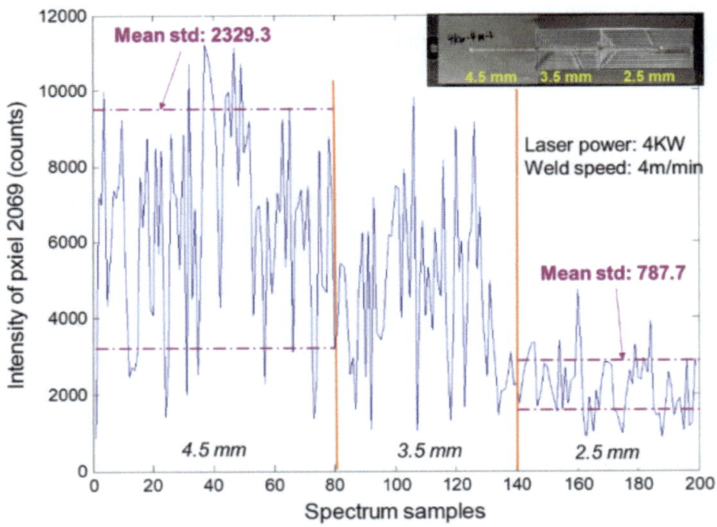

b) Plate thickness

Fig. 4.20 Spectrum intensity of Mg I (517.27 nm) at different parameters of laser welding

(2) Feature correlation evaluation using Pearson CC

The feature correlation between the identified line spectra and elements was quantitatively assessed using the Pearson Correlation Coefficient (CC) method, which is defined as follows:

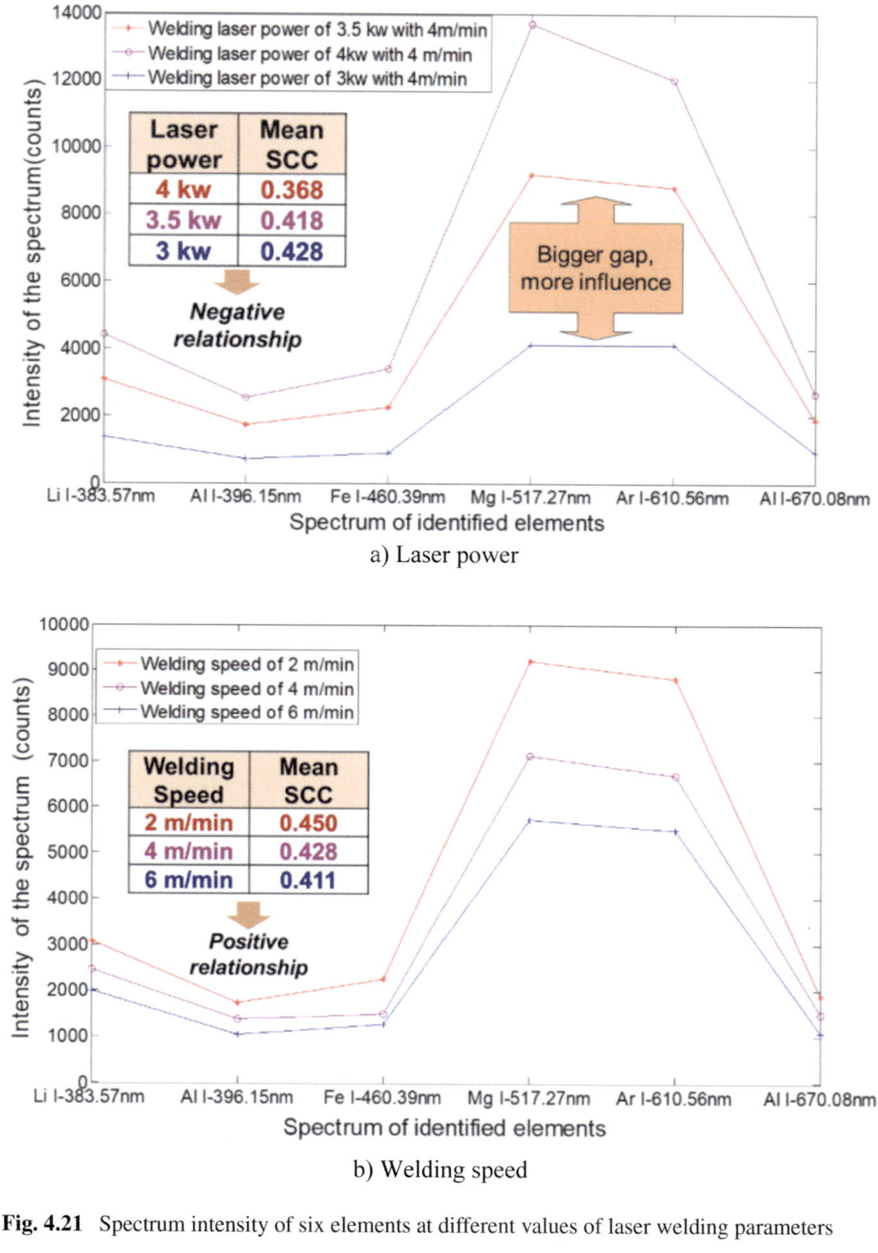

a) Laser power

b) Welding speed

Fig. 4.21 Spectrum intensity of six elements at different values of laser welding parameters

$$r(X,Y) = \frac{\sum_{i=1}^{n}(X_i - \overline{X})(Y_i - \overline{Y})}{\sqrt{\sum_{i=1}^{n}(X_i - \overline{X})^2}\sqrt{\sum_{i=1}^{n}(Y_i - \overline{Y})^2}} \tag{4.10}$$

In the formula, $\overline{X}$ and $\overline{Y}$ represent the mean values of variables X and Y, averaged over n observations, respectively. The Pearson Correlation Coefficient (CC) can also be interpreted as the ratio of the covariance to the standard deviation of the two variables. Another function of the Pearson CC is to assess the extent of linear correlation between X and Y. The Pearson CC was computed for each identified elemental line spectrum against nine other line spectra based on spectral intensity, as detailed in Table 4.3. It was observed that all correlation coefficients exceeded 0.85, with some nearing 0.99. This indicates a substantial presence of abundance and linearity within the feature space, which could impact model performance and thus necessitates reduction and optimization.

(3) Feature extraction of optical spectrum using RFPCA

This section presents the RFPCA (Random Forest Principal Component Analysis) approach, designed for feature dimensionality reduction and optimization. The method unfolds in two primary stages:

1. Feature Selection: From the initial 130-dimensional feature space, 30 spectral pixels are selected based on their Mean Decrease in Gini (MDG) scores derived from the Random Forest (RF) algorithm.
2. Feature Transformation: The chosen pixels are then transformed into a new set of features, termed principal component features, utilizing Principal Component Analysis (PCA).

To assess the importance of spectral features, we employed the Mean Decrease in Gini (MDG) index from the random forest model. This model not only yields classification outcomes but also outputs a feature importance value. The importance

Table 4.3 Pearson CC of the identified element line spectra

	Li I	Al I	Fe I	Li I	Mg I	Mg I	Ar I	Al I	Al II	Fe I
Li I		0.9721	0.9734	0.9472	0.9939	0.9938	0.9865	0.8704	0.9950	0.9261
Al I			0.8768	0.9958	0.9349	0.9876	0.9713	0.8790	0.8964	0.8746
Fe I				0.9170	0.9506	0.9539	0.9505	0.9879	0.8615	0.9652
Li I					0.8810	0.9702	0.9783	0.9631	0.9578	0.9942
Mg I						0.9905	0.9127	0.8610	0.9562	0.9194
Mg I							0.8962	0.8832	0.8630	0.9019
Ar I								0.9112	0.9267	0.8703
Al I									0.9929	0.9663
Al II										0.9492
Fe I										

level of each input feature is ascertained through this value. The reduction in the Gini index is calculated by comparing the index values before and after classification within the sample set. A greater decrease in the Gini index signifies a higher MDG value, which corresponds to a lower impurity in the post-classification sample sets, thereby indicating a feature's heightened importance.

For a set (T) partitioned into (k) classes, the Gini index can be articulated as follows:

$$Gini(T) = 1 - \sum_{j=1}^{k} p(j|T)^2 \tag{4.11}$$

where $p(j|t)$ is the probability of class j at node t.

$$J_b(x_j) = Gini(D) - \sum_{i=1}^{n} \left(\frac{|D_i|}{|D|} \right) Gini(D_i) \tag{4.12}$$

where $J_b(x_j)$ denoted the MDG value of the jth feature;$Gini(D)$

The MDG value of the jth feature is denoted as $J_b(x_j)$; $Gini(D)$ represents the Gini value of the sample set prior to classification; $Gini(D_i)$ signifies the Gini value of the sample subset following classification; and i denotes the class number.

Based on the average SSC value, all samples were bifurcated into two distinct categories using a threshold of 0.4. Subsequently, a RF classification model was constructed to determine the feature importance for each spectral pixel within the 130-dimensional input feature space. The top 30 pixels, as identified by the model, are cataloged in Table 4.3. This table not only enumerates the MDG values and the corresponding wavelengths for each pixel but also suggests their potential chemical constituents.

As delineated in Table 4.4, the paramount MDG value of 4.069 was observed at 395.09 nm, proximate to Li I, succeeded by a value of 3.275 at 669.84 nm, adjacent to Al I. Intriguingly, Mg I at 518.4 nm exhibited a mere MDG value of 1.584. Furthermore, it is noteworthy that most of the line spectra identified are absent from Table 4.4. Conversely, pixels manifesting elevated MDG values predominantly cluster near and around the line spectra.

To enhance the feature set chosen by the Random Forest (RF) algorithm, Principal Component Analysis (PCA) was employed to project the existing features into a new space characterized by reduced correlation and increased variance. In this study, PCA played a crucial role in assessing and selecting the line spectrum through the use of the principal components (PC) vector and its corresponding coefficients. For representing N samples of welding seam conditions with a D-dimensional vector, the eigen structure decomposition was determined as follows:

$$\lambda_i e_i = Q e_i \tag{4.13}$$

Table 4.4 Selected spectrum pixels and elements using the RF

Fea. No	Pix. No	MDG	Wavelength	Fea. No	Pix. No	MDG	Wavelength
1	1023	1.547	382.95 nm	16	1123	1.851	396.01 nm close to Al I
2	1024	1.583	383.08 nm	17	1124	1.553	396.14 nm-Al I
3	1025	1.498	383.21 nm	18	1125	1.625	396.27 nm close to Al I
4	1026	1.902	383.34 nm close to Li I	19	1616	2.691	459.71 nm
5	1027	1.930	383.48 nm close to Li I	20	2068	2.168	517.14 nm close to Mg I
6	1028	1.867	383.61 nm Close to Li I	21	2077	1.508	518.28 nm close to Mg I
7	1029	1.888	383.74 nm close to Li I	22	2078	1.584	518.4 nm-Mg I
8	1109	1.609	394.18 nm	23	2816	2.057	610.03 nm
9	1110	1.472	394.31 nm	24	2818	1.676	610.28 nm
10	1111	1.450	394.44 nm	25	2819	1.642	610.4 nm close to Ar I
11	1112	2.586	394.57 nm	26	2820	1.669	610.52 nm close to Ar I
12	1116	4.069	395.09 nm close to Al I	27	2821	1.930	610.64 nm close to Ar I
13	1120	3.046	395.61 nm	28	3309	3.212	669.72 nm
14	1121	1.885	395.75 nm	29	3310	3.275	669.84 nm close to Al I
15	1122	1.885	395.88 nm	30	3325	2.492	671.64 nm close to Al I

Here, $Q = XX^T$ represents the covariance matrix, and λ_i signifies the eigenvalue corresponding to the eigenvector e_i. The matrix W denotes the linear transformation that projects the original D-dimensional space into a P-dimensional feature subspace. Consequently, a new feature vector Y can be formulated as:

$$Y_j = W^T x_i; \ (j = 1, 2, \ldots, P) \tag{4.14}$$

The matrix W's columns consist of the eigenvectors e_i, which are arranged in descending order based on their associated eigenvalues. The spectrum features preselected and detailed in Table 4.2 underwent PCA processing, effectively reducing the dimensionality from the original 30 (denoted as RF-30D_org) to 25 (denoted as RFPCA-25D). To evaluate the efficacy of various feature subsets derived through

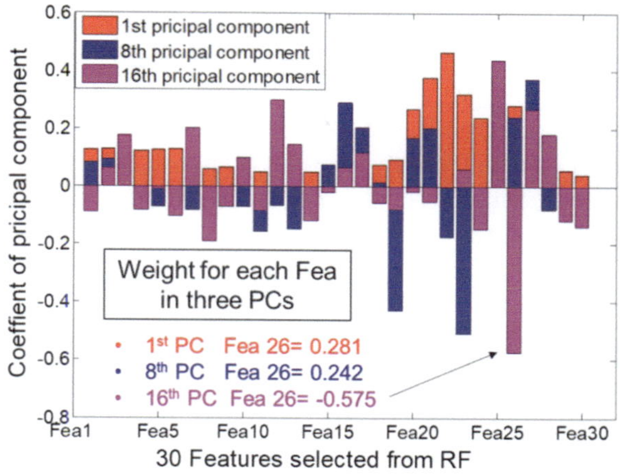

Fig. 4.22 The results of three cases of the PCA performed on RF features

RFPCA, two additional feature sets were extracted: the top ten dimensions (RFPCA-F10D) and the bottom ten dimensions (RFPCA-L10D) of the RFPCA. Furthermore, fast Independent Component Analysis (fast-ICA) was applied to diminish and reconstitute the feature space of the optical spectrum signal, resulting in the RFFICA-10D, which is recognized as a non-linear dimensionality reduction technique. This facilitated a more profound comparison between the original signal and those reconstructed by PCA and ICA.

The detailed results of Principal Component Analysis (PCA) for three cases of new principal component (PC) features are presented in Fig. 4.22. The y-axis corresponds to the eigenvector e_i, which represents the coefficient of each principal component.

In this context, the eigenvector e_i, can be interpreted as the weight assigned to each original feature (Fea1, Fea2, ..., Fea30) when linearly composing the new spectrum feature and transforming it into the principal component. A higher value of the eigenvector e_i indicates greater importance for the corresponding feature. Figure 4.22 displays the first, eighth, and sixteenth PCs along with their corresponding eigenvector e_i values for the preselected 30 pixels of the spectrum. Notably, Fea26 exhibits different weights: 0.281 in the first PC and -0.575 in the sixteenth PC. This variability contributes to PCA's ability to enhance diversity and reduce redundancy in the feature space.

Furthermore, Fea26 (610.52 nm, close to Ar I), which has an MDG value of 1.669 (as shown in Table 4.4), demonstrates a higher coefficient value in Fig. 4.22. Similarly, Fea22 (518.4 nm, associated with Mg I) displays a lower MDG value in Table 4.4 but the highest coefficient value in the first PC in Fig. 4.22. Thus, the MDG importance from the Random Forest (RF) and the coefficient weights from PCA effectively evaluate feature importance from different perspectives, showcasing certain complementarity.

In summary, the combination of RF and PCA allows us to uncover hidden information within the high-dimensional spectrum feature.

To assess the linear correlation of the spectral features before and after the application of PCA to the RF features, the Pearson Correlation Coefficient (CC) values and scatter plots for both the original spectral features and the RFPCA-derived features are depicted in Fig. 4.23. It is evident that the high degree of linear correlation present in the original feature space has been markedly diminished, from a CC value of 0.91 to 0.38. This substantial reduction underscores the efficacy of the RFPCA method and significantly enhances the diversity of the feature space.

4.3.3 Spectral Data Analysis Process Based on RFPCA-XGBoost Method

The following section outlines the proposed RFPCA-XGBoost regression model, which is based on ensemble learning and comprises two main components: feature extraction and selection as detailed in Sect. 4.2, and the tree-based regression model of XGBoost.

The tree-based models, as shown in Fig. 4.24, are constructed based on the concept of Decision Tree (DT), which aims to identify the most homogeneous partition referred to as purity. The ID3 algorithm uses information gain to measure impurity, while the C4.5 algorithm utilizes information gain rate and the CART (Classification And Regression Tree) algorithm employs Gini coefficient for this purpose. Despite being widely used due to its fast computation speed, high efficiency, and ease of interpretation, the single-classification decision classifier of DT restricts its application scenarios and scopes, resulting in limited robustness and susceptibility to overfitting.

The Random Forest (RF) proposed by Breiman [35] in 2001, integrates the concept of bagging into ensemble learning, as illustrated in Fig. 4.24, which extends the sample space through bagging and constructs multiple CART decision trees. Bagging can be simply understood as: bootstrap aggregating for sampling, majority voting for classification, or simple averaging for regression. Furthermore, the base learners of bagging are generated simultaneously without strong interdependence among them.

Friedman [36] proposed the Gradient Boosting Decision Tree (GBDT) in 2002, which utilizes boosting to derive a new base learner focused on reducing misclassified data and residual error rather than solely enhancing classification accuracy. The boosting tree model can be depicted as an additional model with CART tree as the base learner, utilizing the Gini coefficient for measuring impurity.

$$f\left(\overrightarrow{X}\right) = f_M\left(\overrightarrow{X}\right) = \sum_{m=1}^{M} h_m\left(\overrightarrow{X};\ \Theta_m\right) \tag{4.15}$$

The symbol $h_m(\overrightarrow{X};\ \Theta_m)$ represents the mth CART tree, while Θ_m represents the parameters of the mth CART tree. The square error loss function of the GBDT can be defined as follows:

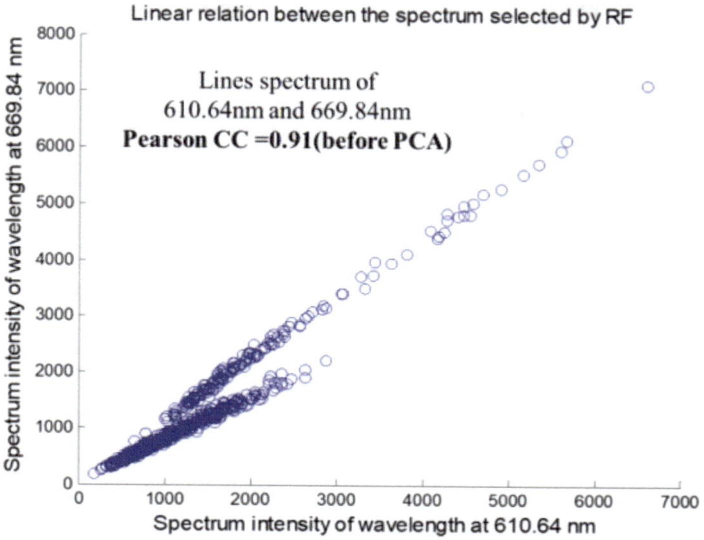

a) Scatter of two line spectra

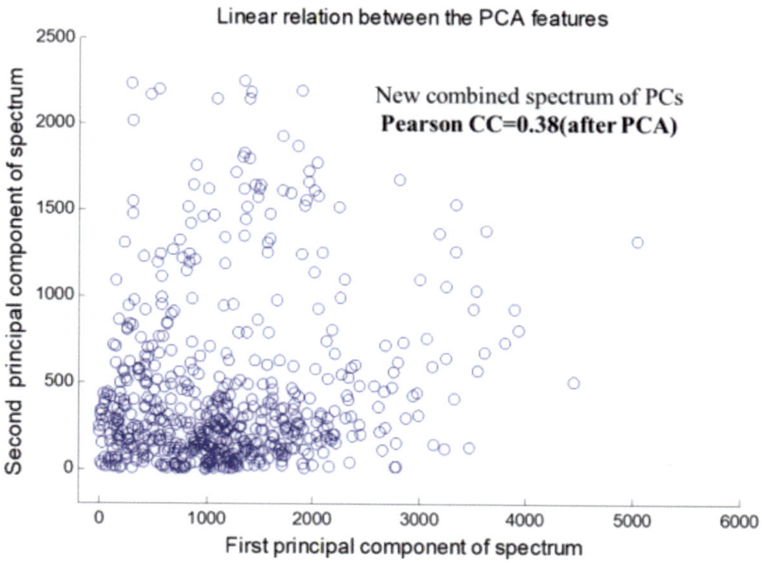

b) Scatter of the first and second PCs

Fig. 4.23 The Pearson CC and scatter images of the original spectrum and RFPCA feature

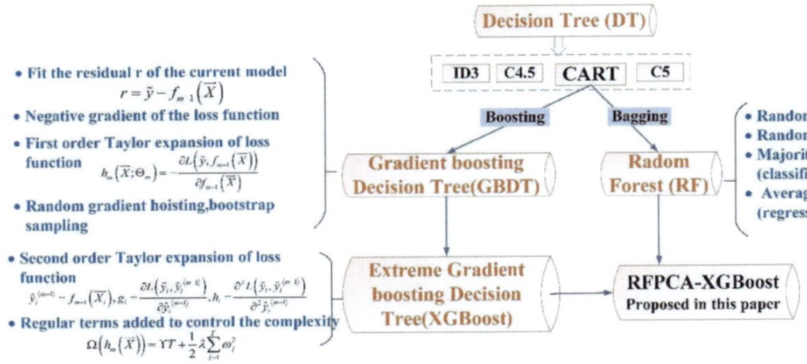

Fig. 4.24 Family tree of typical tree-based models in ensemble learning

$$
\begin{aligned}
L\left(\widetilde{y}, f_m\left(\overrightarrow{X}\right)\right) &= L\left(\widetilde{y}, f_m\left(\overrightarrow{X}\right) + h_m\left(\overrightarrow{X};\, \Theta_m\right)\right) \\
&= \left(\widetilde{y} - f_{m-1}\left(\overrightarrow{X}\right) - h_m\left(\overrightarrow{X};\, \Theta_m\right)^2\right) \\
&= \left(r - h_m\left(\overrightarrow{X};\, \Theta_m\right)\right)^2
\end{aligned}
\tag{4.16}
$$

The boosting tree algorithm trains each individual tree to fit the residuals of the current model, with $r = \widetilde{y} - f_{m-1}\left(\overrightarrow{X}\right)$ representing these residuals. In GBDT, an approximation for the residual is obtained by using the negative gradient of the loss function, effectively serving as a first-order Taylor expansion of the loss function.

$$
\begin{aligned}
\Delta L &= L\left(\widetilde{y}, f_m\left(\overrightarrow{X}\right)\right) - L\left(\widetilde{y}, f_{m-1}\left(\overrightarrow{X}\right)\right) \\
&= \frac{\partial L\left(\widetilde{y}, f_{m-1}\left(\overrightarrow{X}\right)\right)}{\partial f_{m-1}\left(\overrightarrow{X}\right)} h_m\left(\overrightarrow{X};\, \Theta_m\right)
\end{aligned}
\tag{4.17}
$$

The derivative of the loss function is as follows:

$$
h_m\left(\overrightarrow{X};\, \Theta_m\right) = -\frac{\partial L\left(\widetilde{y}, f_{m-1}\left(\overrightarrow{X}\right)\right)}{\partial f_{m-1}\left(\overrightarrow{X}\right)}
\tag{4.18}
$$

In contrast, the Random Forest (RF) employs fully developed sub-decision trees (low bias, high variance) and emphasizes the independence of subtrees. Meanwhile, in Gradient Boosting Decision Trees (GBDT), each base learner is assigned a different weight representing its success in the previous iteration.

In the optimization process, GBDT utilizes solely the first derivative information, while XGBoost makes use of the second-order Taylor expansion of the cost function to acquire both first and second derivatives. Furthermore, XGBoost displays superior efficiency in numerical optimization when compared to GBDT and is notably more complex with regards to loss functions (the error between predicted and true values) as shown in Figs. 4.24 and 4.25 respectively.

It is crucial to acquire the best tree structure for each sample and base learner, ensuring minimization of the global target function. The utilization of a greedy algorithm is imperative, and it is outlined as follows:

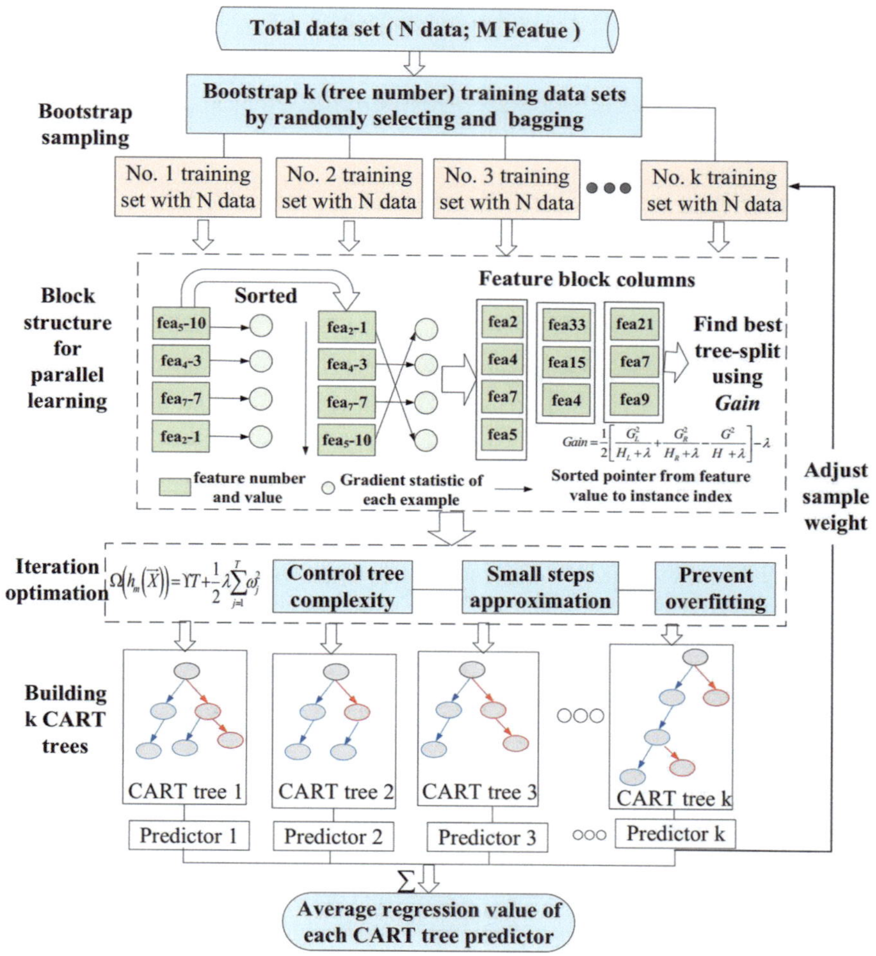

Fig. 4.25 Diagram of the XGBoost regression model

$$\begin{aligned} G_L &= \sum_{i\in\|_L} g_i,\ G_R \sum_{i\in\|_R} g_i, \\ H_L &= \sum_{i\in\|_L} h_i,\ H_R \sum_{i\in\|_R} h_i, \\ G &= \sum_{i\in\|_L} g_i + \sum_{i\in\|_R} g_i = G_L + G_R \\ H &= \sum_{i\in\|_L} h_i + \sum_{i\in\|_R} h_i = H_L + H_R \end{aligned} \tag{4.19}$$

Gj represents the sum of first-order partial derivatives for samples in the left leaf nodes, and *Hj* represents the sum of first-order partial derivatives for samples in the right leaf nodes.

The splitting gain of leaf nodes can be defined as follows:

$$Gain = \frac{1}{2}\left[\frac{G_L^2}{H_L + \lambda} + \frac{G_R^2}{H_R + \lambda} - \frac{G^2}{H + \lambda}\right] - \lambda \tag{4.20}$$

The left subtree of the leaf node is represented by $\frac{G_L^2}{H_L+\lambda}$, while $\frac{G_R^2}{H_R+\lambda}$ represents the right subtree. $\frac{G^2}{H+\lambda}$ denotes the structure of the leaf node itself when it remains unsplit. If Gain exceeds 0, then the leaf may or may not split. The greedy algorithm exhaustively explores all features and potential split-points in order to identify the optimal one. In Table 4.7, an exhaustive scan was performed for all possible splitting points within each leaf node as part of the exact greedy algorithm. Subsequently, the gain associated with each splitter point was computed and the one with maximum gain was selected as that particular leaf node's optimal splitter point. This approach can lead to significant computational expenses. To mitigate this issue, XGBoost employs column sampling similar to RF, which also serves to prevent overfitting. A detailed diagram is provided in Fig. 4.25 (Tables 4.5 and 4.6).

The proposed RFPCA-XGBoost algorithm offers several advantages over other tree-based models, including the ability to quantitatively evaluate feature importance using MDG in the RF, improve feature diversity through PCA, prevent overfitting, and better control the complexity of the tree algorithm and block structure for parallel learning in XGBoost.

The XGBoost algorithm employs the reduction of structural risk to determine the parameters of the subsequent decision tree through a forward-stepping approach.

$$\Theta_m = \arg\min_{\Theta_m} \sum_{i=1}^{N} L\left(\tilde{y}_i, f_m\left(\vec{X}\right)\right) + \Omega\left(h_m\left(\vec{X}\right)\right) \tag{4.21}$$

The regularization term of the mth decision tree is denoted by $\Omega(h_m(\vec{X}))$. The second-order Taylor expansion of the loss function is provided.

Table 4.5 Evaluation criteria of the RFPCA-XGBoost regression model

Evaluation criteria	Formula	Description
Mean squared error (MSE)	$\overline{y}_m = \frac{1}{N_m}\sum_{i\in N_m} y_i$ $MSE(X_m) = \frac{1}{N_m}\sum_{i\in N_m}\left(y_i - \overline{y}_m\right)^2$	X_m denotes the training or test data in tree-node m
Mean absolute error (MAE)	$\overline{y}_m = \frac{1}{N_m}\sum_{i\in N_m} y_i$ $MAE(X_m) = \frac{1}{N_m}\sum_{i\in N_m}\left\lvert y_i - \overline{y}_m\right\rvert$	$\hat{y}_m$ denotes the predicted value, and $\overline{y}_m$ is the mean value. The molecular part is similar to the MSE, while the numerator is more like the variance
Determinate coefficient R-Square (R^2)	$\overline{y}_m = \frac{1}{N_m}\sum_{i\in N_m} y_i$ $R^2 = 1 - \frac{\sum_{i\in N_m}(y_i - \hat{y}_m)^2}{\sum_{i\in N_m}\left(y_i - \overline{y}_m\right)^2}$	The value range of R^2 is [0, 1]. The higher R^2 is, the better the regression model will be

Table 4.6 Ranges and fixed-parameter values of the RFPCA-XGBoost prediction model

Parameter	Description	Grid search range
Tree-estimators	Number of boosted trees	(10–600)
Max_depth	Maximum tree depth for base learners	(1–20)
Gamma	Minimum loss reduction	(0–0.4)
Learning_rate	Boosting learning rate	(0.01–0.2)
Subsample	Subsample ratio of the training instance	(0.05–1)
Parameter	*Description*	*Fixed value*
Booster	Booster to use	'gbtree'
Objective	The loss function needs to be minimized	Gamma regression
Min child	Minimum sum of instance weight	1
Weight	(hessian) needed in a child	
Lambda	L2 regularization term on weights	3
Colsample by	Subsample ratio of columns when	0.6
Tree	Constructing each tree	
Seed	Random number seed	90
tree_method	The tree construction algorithm used in XGBoost	Exact greedy
Other parameters use the default value		

$$\hat{y}_i^{\langle m-1\rangle} = f_{m-1}\left(\overrightarrow{X}\right), g_i = \frac{\left(\partial L\hat{y}_i, \hat{y}_i^{\langle m-1\rangle}\right)}{\partial \hat{y}_i^{\langle m-1\rangle}}, h_i = \frac{\partial^2 L\left(\hat{y}_i, \hat{y}_i^{\langle m-1\rangle}\right)}{\partial^2 \hat{y}_i^{\langle m-1\rangle}} \tag{4.22}$$

Table 4.7 Comparison results of XGBoost prediction models

Regression model	Mean			Std		
Evaluation index	MSE	R^2	MAE	MSE	R^2	MAE
RF-XGBoost	0.0031	0.2947	0.0427	0.0004	0.0966	0.0026
RFFICA-XGBoost	0.0006	0.8807	0.0135	0.0002	0.0393	0.0026
RFPCA-XGBoost	0.0003	0.9383	0.0104	0.0001	0.0208	0.0014

The derivatives of the loss function g_i to h_i are represented as $L\left(\widetilde{y}_i, \hat{y}_i^{\langle m-1\rangle}\right)$ and $\hat{y}_i^{\langle m-1\rangle}$, respectively. Subsequently, the second-order Taylor expansion of the loss function can be expressed as:

$$\begin{aligned} L &= \sum_{i=1}^{N} L\left(\widetilde{y}, f_m\left(\vec{X}\right)\right) + \Omega\left(h_m(\vec{X})\right) \\ &= \sum_{i=1}^{N} L(\widetilde{y}_i, \hat{y}_i^{\langle m-1\rangle} + h_m\left(\vec{X}_i\right) + \Omega\left(h_m\left(\vec{X}\right)\right) \\ &\simeq \sum_{i=1}^{N}\left[L(\widetilde{y}_i, \hat{y}_i^{\langle m-1\rangle} + g_i h_m\left(\vec{X}_i\right) + \frac{1}{2} h_i h_m^2\left(\vec{X}_i\right)\right] \\ &\quad + \Omega\left(h_m\left(\vec{X}\right)\right) + \text{constant} \end{aligned} \tag{4.23}$$

The regularization term is composed of two components:

$$\Omega\left(h_m\left(\vec{X}\right)\right) = \Upsilon T + \frac{1}{2}\lambda \sum_{j=1}^{T} \omega_j^2 \tag{4.24}$$

This section demonstrates the complexity of the decision tree, where T is used to represent the number of leaf nodes, ωj denotes the output value associated with each leaf node, and γ, λ represents the coefficient that controls the proportion of each part. As the quantity of tree nodes increases, so does the intricacy of DT. This results in a decrease in model variance, leading to simplification of the learned model and prevention of overfitting. This is one reason for XGBoost's superior performance compared to traditional GBDT. The proposed RFPCA-XGBoost algorithm for predicting SSC in laser welding can be found in Algorithm 1.

Algorithm 1: Proposed RFPCA-XGBoost for SSC prediction
• **Input**
Training data set: 3648 dimension spectrum data of laser welding seam
Loss function L:$\sum_{i=1}^{N}\left[L(\widetilde{y}_i, \hat{y}_i^{\langle m-1\rangle} + g_i h_m\left(\vec{X}_i\right) + \frac{1}{2} h_i h_m^2\left(\vec{X}_i\right)\right]$

(continued)

(continued)

• **Output**: Regression results of SSC	
for $i = 1$ to m (m = data blocks)	
	• Identify lines peaks of spectrum data and reduce down to 130 dimension
	• Reduce down to 30 dimension based on RF classification model and feature importance of Gini index $Gini(T) = 1 - \sum_{j=1}^{k} p(j\|T)^2$
	• PCA feature transformation an dimension reduction and obtain 25 dimension of new PCs
	• In itialization of XGBoost obj: gamma regression with log-link; base_score: 0.5; eval_metric: negative log-likelihood for gamma regression
	• Bootstrap sampling and randomly select feature set, calculate the loss function value: $\sum_{i=1}^{N}\left[L(\tilde{y}_i, \hat{y}_i^{(m-1)}) + g_i h_m\left(\vec{X}_i\right) + \frac{1}{2} h_i h_m^2\left(\vec{X}_i\right)\right] + \Omega\left(h_m\left(\vec{X}\right)\right)$
	• Find the best tree splitting with best feature set based on nodes gain and greedy algorithm $Gain = \frac{1}{2}\left[\frac{G_L^2}{H_L+\lambda} + \frac{G_R^2}{H_R+\lambda} - \frac{G^2}{H+\lambda}\right] - \lambda$
	• Calculate the average regression value of each CART tree
	• Update the current function value
end	

The MSE, MAE, and R^2 were utilized as evaluation metrics to assess the performance of the RFPCA-XGBoost regression model from various perspectives. The results are presented in Table 4.5.

4.3.4 Discussion on Grid Parameters, Different Models and Feature Sets

The proposed RFPCA-XGBoost is compared with RF, GBDT, DT, and LR using different input feature sets and feature dimensions of the original spectrum, PCA, and fast ICA in this section. Additionally, the regression tree model's mechanism is analyzed through visualizing the tree splitting process before discussing potential applications and future work.

The PyCharm 2019.1.3 software was utilized on an Intel Core i7-6560U CPU operating at a frequency of 2.20 GHz and equipped with 16 GB of RAM. The tree estimator and tree max_depth was recognized as the two crucial parameters in the tree-based regression model. As illustrated in Fig. 4.26a and b, RFPCA-XGBoost demonstrated superior R^2 values for the two key parameters during a stable phase when compared to other tree-based models and LR. Specifically, with an increase in the number of tree estimators, different tree-based SSC regression models exhibited diverse patterns of change. In Fig. 4.26a, as the size of the tree estimator increased from 10 to 110, R^2 for the XGBoost model showed the most substantial growth, while others displayed comparatively less advancement, particularly RF and DT.

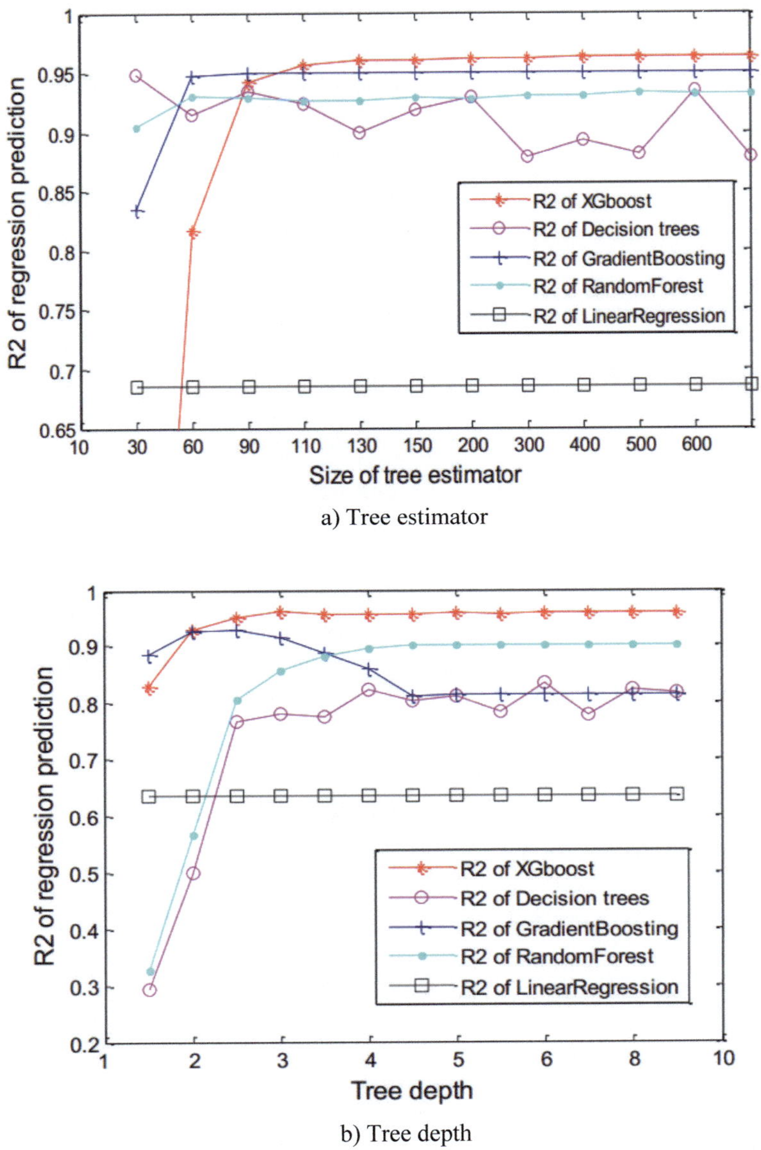

Fig. 4.26 The R^2 values of the SSC regression models (RFPCA-XGBoost, DT, GB, RF, and LR) at different parameters

Based on the data presented in Fig. 4.26a, it is evident that for tree estimator sizes exceeding 110, most models exhibited stable performance, with the exception of the Decision Tree (DT). Specifically, while the R^2 value for DT fluctuated significantly from 0.85 to 0.95, other tree models maintained a relatively stable R^2 around 0.95. In comparison to the tree estimator and as depicted in Fig. 4.26b, it can be observed that tree depth had a lesser impact on the R^2 of the XGBoost model, whereas DT and RF were more affected by smaller tree depths. Notably, overfitting was detected in Gradient Boosting (GB) when the tree depth exceeded four.

The best model performance was achieved at a gamma value of zero in Fig. 4.27a, which was the default setting. A higher gamma resulted in a more cautious algorithm. As illustrated in Fig. 4.27b, a reduced learning rate led to increased R^2 and decreased MSE. Furthermore, XGBoost would randomly select a specific portion of the training data before growing trees, and an inappropriate sample size could result in under-fitting or over-fitting of the regression prediction model. The findings presented in Fig. 4.27c suggest that the subsample ratio should exceed 0.5. Additionally, smaller increments were utilized during grid search, such as 0.7, 0.73, 0.76, and 0.8 as evidenced in Fig. 4.27c. Using more than seventy percent of the training data might produce superior regression trees and prevent overfitting.

After conducting ten rounds of testing, the XGBoost model achieved the highest mean R^2 value of 0.9383 for the input feature set RFPCA-25D, followed by RF (0.9062), GBDT (0.8767), and DT (0.8456), while LR had the lowest mean value of 0.6179. The performance of all five models was significantly enhanced by RFPCA-25D for different input feature sets.

The total number of data samples was 583 with a subsample ratio set to 0.76 based on optimal grid search results, indicating that 76% of the entire dataset was randomly sampled for training while the remaining data was used for testing.

The tree-based models—XGBoost, DT, RF, and GBDT—were configured with identical values for tree depth and tree estimators according to grid search results: max-depth set to 8; estimator set to 300; learning rate set to 0.05; gamma set to zero within XGBoost model parameters.

Each model underwent ten rounds of training and testing without setting random_state for data splitting to zero before calculating mean and standard deviation values for MSE, R^2, and MAE (Fig. 4.28).

The comparison results of the XGBoost prediction models for different feature sets are presented in Table 4.7, demonstrating a significant increase in the mean R^2 value by approximately 68.6% from 0.2947 for RF-XGBoost to 0.9383 for RFPCA-XGBoost, while exhibiting the smallest standard deviation of 0.0208.

On the other hand, the RFFICA-XGBoost surprisingly demonstrated a reduced R^2 value of 0.8807, despite its robust processing capability in non-linear situations. Furthermore, the RFPCA-L10D model outperformed the RFPCA-F10D model by almost double across all five feature sets. This could be due to the approach of averaging SSC values for each section of the welding seam, as detailed in Sect. 1. 2, which requires minor adjustments to the original data length. As is commonly understood, PCA is adept at capturing information with high variance and rate of

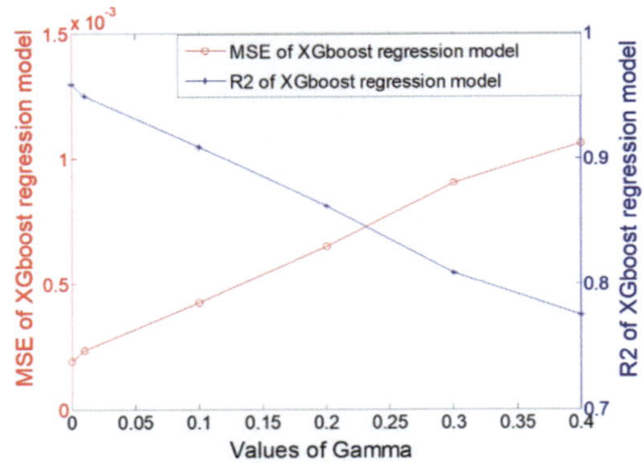

a) Gamma value

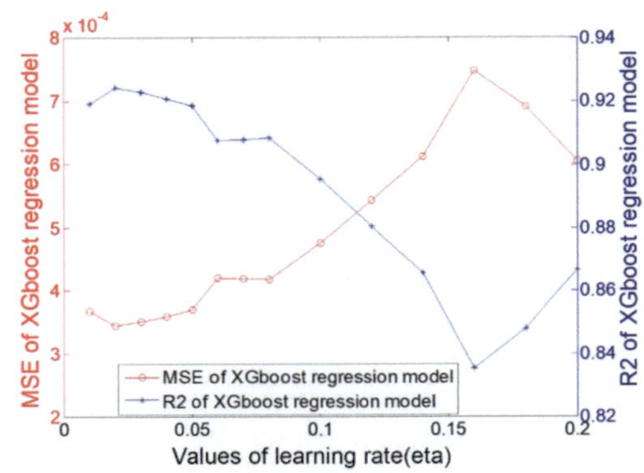

b) Learning rate

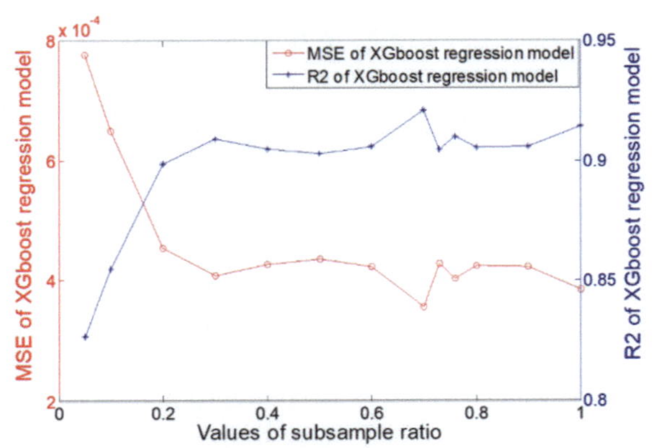

c) Subsample ratio

Fig. 4.27 The R^2 and MSE values of the RFPCA-XGBoost regression model at different parameters

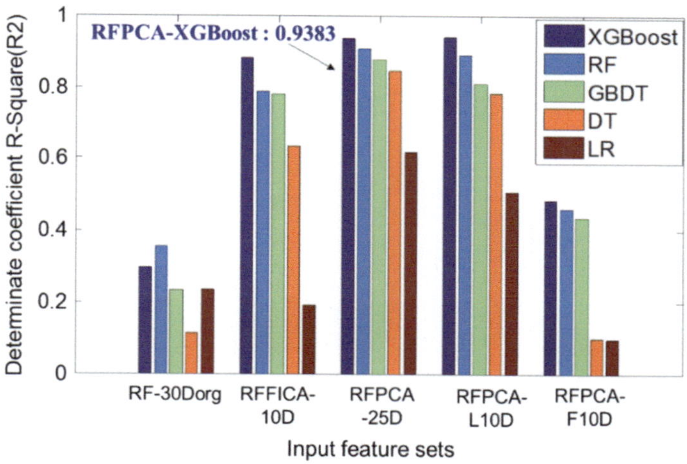

Fig. 4.28 Comparison results of different tree-based models and feature sets

change; thus, the latter principal components contain less variance and are more indicative of average SSC values.

Besides, as presented in Fig. 4.29a, the mean relative error of the RF-XGBoost model was 10.17%, while it was only 2.42% for the proposed model, the RFPCA-XGBoost, thus proving better performance than other combinations and tree-based models.

The visualization of the proposed regression model is presented in Fig. 4.30, where detailed parameters and regression results are provided. In Fig. 4.30, f denotes the feature number within the feature set. In this study, features No. 21 (root node), No. 14, and No. 16 (child node) were selected for decision making. Analysis of the scatter plot in the upper left corner of Fig. 4.28 reveals that a lower R^2 value of 0.774 was obtained when gamma was set to 0.4.

Furthermore, as shown in Fig. 4.30, a decision tree with fewer root nodes, child nodes, tree leaves, and shallower depth demonstrates more accurate performance prediction compared to a deeper tree model.

4.3.5 Conclusions

The relationship between the Mg I spectrum at 517.27 nm and the average SCC is depicted in Fig. 4.31. The mean spectral intensity of Mg I was calculated within the corresponding area of the test specimen. Based on the outcome of cubic spline interpolation, a significant linear and positive association between Mg I spectrum intensity and SSC is evident, laying a crucial groundwork for predicting and regressing seam strength.

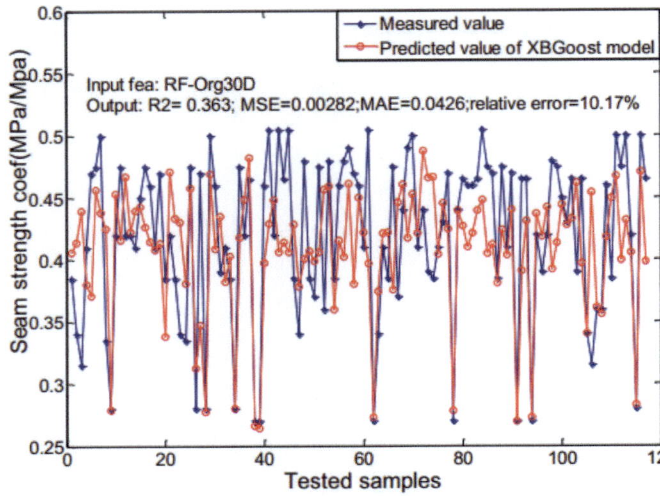

a) RF-XGBoost

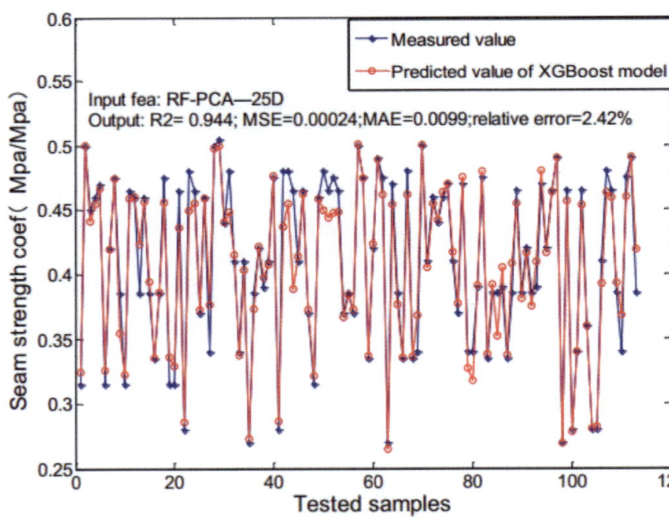

b) RFPCA-XGBoost

Fig. 4.29 Test results of different regression models

The results of feature selection using different methods and their respective selection principles are detailed in Table 4.8. Upon comparison of the outcomes in Fig. 4.31 and Table 4.8, it becomes apparent that Fea. 26 (610.52 nm-close to Ar I), Fea. 22 (518.4 nm-Mg I) and other features demonstrate varying degrees of importance when assessed by PCA and RF, suggesting that the MDG importance of RF and the coefficient weight of PCA can effectively evaluate the significance of these features from

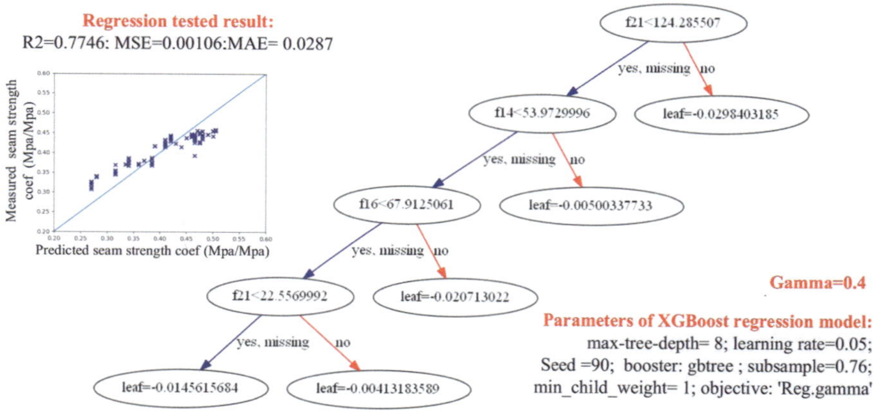

Fig. 4.30 Visualization of the RFPCA-XGBoost regression model

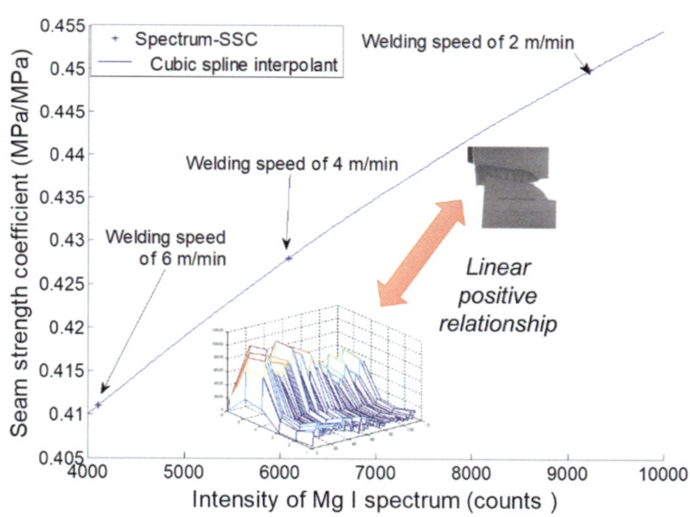

Fig. 4.31 Linear relationship between the spectrum and SSC at different welding speeds

diverse perspectives. Consequently, through the integration of these two methodologies, the proposed RFPCA exhibits strong complementarity; specifically, PCA is capable of extracting data-driven features influenced by dispersion degree, while RF is driven by classification objectives and internal mechanisms. The RFPCA has the potential to reveal additional latent information within high-dimensional spectrum features.

This work identifies ten line spectra and elements, including Li, Al, and Mg, in the laser welding process of 2195 Al-Li alloy. It is important to note that chemical elements may exhibit variations for different welded materials, depending on the base material components and ionization conditions. However, the proposed method

Table 4.8 Comparison of the feature selection methods

Feature selection method	PCA	RF
Evaluation index	Coefficient weight of new PC	Mean decrease in Gini (MDG)
Selected feature of spectrum	Fea. 22 (518.4 nm-Mg I) Fea 0.26 (610.52 nm-close to Ar I)	Fea. 12 (395.06 nm-close to Li I) Fea. 29 (669.84 nm-close to Al I)
Selection principle	Driven by data itself and dispersion degree	Driven by classification purpose and mechanism
Proposed RFPCA	Evaluate feature importance from different aspects good complementarity	

demonstrates good generality as the RFPCA automatically selects the lines spectrum based on its feature importance index. Furthermore, using the proposed RFPCA results in a higher degree of feature diversity compared to the original feature space method, proving its effectiveness. The dimension of the RFPCA feature space shows that the new PC features have greater importance than others, consistent with PCA's contribution rate.

The work has the following conclusions:

1. Introducing a new feature extraction and regression model, RFPCA-XGBoost, for predicting SSC regression of 2195 Al-Li alloy during laser welding. This model holds promise for generalization to diverse materials like titanium alloy and nickel alloy. Furthermore, it is applicable to a wide range of fusion welding manufacturing processes that encompass arc and light sources, such as TIG, MIG, CMT, WAAM, and laser additive manufacturing.
2. The quantitative evaluation of the line spectrum's importance reveals that the highest value (4.069) is observed at 395.09 nm (close to Li I), followed by 3.275 at 669.84 nm (close to Al I), and 1.584 for Mg I at 518.4 nm. In contrast, the Ar I line spectrum demonstrates lower importance. The proposed RFPCA exhibits strong complementarity and has the capability to unveil hidden information within high-dimensional spectrum features.
3. According to the regression evaluation criteria of MSE, MAE and R^2, the feature set of RFPCA-25D demonstrates superior regression performance compared to RFPCA-F10D, RFPCA-L10D and the original spectrum data in sequence. Additionally, PCA exhibits better capability than fast-ICA in transforming new feature sets of line spectrum.
4. Among the tree-based ensemble learning models, XGBoost outperforms RF, GBT, DT, and LR in terms of MSE, MAE, and R^2 evaluation criteria. The mean R^2 values for these models are 0.9383, 0.9062, 0.8767, 0.8456, and 0.6179 respectively.

In the future, further in-depth research will be conducted on XGBoost with regards to its new loss function, application in classification problems, and comparison with other boosting algorithms.

As for the optical spectrum of laser beam welding, additional investigation will be carried out on the burning loss mechanism of key trace alloy elements for Al-Li alloy, its correlation with the optical emission pattern, and its impact on seam strength.

References

1. S. B. Chen, *Intelligentized methodology for arc welding dynamical processes: visual information acquiring, knowledge modeling and intelligent control*. Berlin: Springer, 2008.
2. S. Shankar, P. Vilaça, P. Dash, S. Chattopadhyaya, and S. Hloch, "Joint strength evaluation of friction stir welded Al-Cu dissimilar alloys," *Measurement,* vol. 146, pp. 892–902, 11/01/ November 2019 2019, https://doi.org/10.1016/j.measurement.2019.07.019.
3. H. Huang, Z. Feng, N. Ma, and X. Yin, "Finite element analysis and in-situ measurement of out-of-plane distortion in thin plate TIG welding," (in English), *Materials,* Article vol. 12, no. 1, 01/03/ 2019, https://doi.org/10.3390/ma12010141.
4. Z. Chen, J. Chen, and Z. Feng, "Monitoring Weld Pool Surface and Penetration Using Reversed Electrode Images," *WELDING JOURNAL,* vol. 96, no. 10, pp. 367S-375S, 10/01/ 2017.
5. Y.-K. Liu and Y.-M. Zhang, "Supervised learning of human welder behaviors for intelligent robotic welding," *IEEE Transactions on Automation Science and Engineering,* vol. 14, no. 3, pp. 1532–1541, 2015.
6. Y. Liu and Y. Zhang, "Toward welding robot with human knowledge: A remotely-controlled approach," *IEEE Transactions on Automation Science and Engineering,* vol. 12, no. 2, pp. 769–774, 2014.
7. L. Yang, J. Fan, Y. Liu, E. Li, J. Peng, and Z. Liang, "Automatic detection and location of weld beads with deep convolutional neural networks," *IEEE Transactions on Instrumentation and Measurement,* vol. 70, pp. 1–12, 2020.
8. Z. Zhang, G. Wen, and S. Chen, "Audible Sound-Based Intelligent Evaluation for Aluminum Alloy in Robotic Pulsed GTAW: Mechanism, Feature Selection, and Defect Detection," *IEEE Transactions on Industrial Informatics, Industrial Informatics, IEEE Transactions on, IEEE Trans. Ind. Inf.,* Periodical vol. 14, no. 7, pp. 2973–2983, 07/01/ 2018, https://doi.org/10.1109/TII.2017.2775218.
9. H. Gu and W. W. Duley, "A statistical approach to acoustic monitoring of laser welding," (in English), *Journal of Physics D: Applied Physics,* Article vol. 29, no. 3, pp. 556–560, 03/14/ 1996, https://doi.org/10.1088/0022-3727/29/3/011.
10. Z. Zhang, H. Chen, Y. Xu, J. Zhong, N. Lv, and S. Chen, "Multisensor-based real-time quality monitoring by means of feature extraction, selection and modeling for Al alloy in arc welding," *Mechanical systems and signal processing,* vol. 60, pp. 151–165, 2015.
11. Y. Cui, Y. Shi, T. Zhu, and S. Cui, "Welding penetration recognition based on arc sound and electrical signals in K-TIG welding," *Measurement,* vol. 163, 10/15/October 2020 2020, https://doi.org/10.1016/j.measurement.2020.107966.
12. L. Mazzoleni, A. G. Demir, L. Caprio, M. Pacher, and B. Previtali, "Real-time observation of melt pool in selective laser melting: spatial, temporal, and wavelength resolution criteria," *IEEE Transactions on Instrumentation and Measurement,* vol. 69, no. 4, pp. 1179–1190, 2019.
13. S. Jinachandran *et al.*, "Cold crack monitoring and localization in welding using fiber Bragg grating sensors," *IEEE Transactions on Instrumentation and Measurement,* vol. 69, no. 11, pp. 9228–9236, 2020.
14. M. Ferrara, A. Ancona, P. M. Lugara, and M. Sibilano, "Online quality monitoring of welding processes by means of plasma optical spectroscopy," 2000, vol. 3888: SPIE, pp. 750–758.

15. A. Ancona, V. Spagnolo, P. M. Lugarà, and M. Ferrara, "Optical sensor for real-time monitoring of CO_2 laser welding process," *Applied Optics,* vol. 40, no. 33, pp. 6019–6025, 2001/11/20 2001, https://doi.org/10.1364/AO.40.006019.
16. J. Mirapeix, P. B. García-Allende, A. Cobo, O. M. Conde, and J. M. López-Higuera, "Real-time arc-welding defect detection and classification with principal component analysis and artificial neural networks," *NDT & E International,* vol. 40, no. 4, pp. 315–323, 2007/06/01/ 2007, https://doi.org/10.1016/j.ndteint.2006.12.001.
17. J. Mirapeix, E. Vila, J. J. Valdiande, A. Riquelme, M. Garcia, and A. Cobo, "Real-time detection of the aluminium contribution during laser welding of Usibor1500 tailor-welded blanks," *Journal of Materials Processing Technology,* vol. 235, pp. 106–113, 2016/09/01/ 2016, https://doi.org/10.1016/j.jmatprotec.2016.04.017.
18. Y. Huang, S. Hou, T. Sun, L. Yang, S. Li, and S. Chen, "An improved model of porosity formation during pulsed GTA welding of aluminum alloys," *Materials Science and Engineering: B,* vol. 238–239, pp. 122–129, 2018/12/01/2018, https://doi.org/10.1016/j.mseb.2018.11.012.
19. J. Shin and J. Mazumder, "Composition monitoring using plasma diagnostics during direct metal deposition (DMD) process," *Optics & Laser Technology,* vol. 106, pp. 40–46, 10/01 2018, https://doi.org/10.1016/j.optlastec.2018.03.020.
20. L. Song, W. Huang, X. Han, and J. Mazumder, "Real-Time Composition Monitoring Using Support Vector Regression of Laser-Induced Plasma for Laser Additive Manufacturing," *IEEE Transactions on Industrial Electronics,* vol. 64, no. 1, pp. 633-642, 2017, https://doi.org/10.1109/TIE.2016.2608318.
21. Z. Zhang, Z. Yang, W. Ren, and G. Wen, "Random forest-based real-time defect detection of Al alloy in robotic arc welding using optical spectrum," *Journal of Manufacturing Processes,* vol. 42, pp. 51–59, 2019/06/01/ 2019, https://doi.org/10.1016/j.jmapro.2019.04.023.
22. Z. Zhang, G. Wen, and S. Chen, "Weld image deep learning-based on-line defects detection using convolutional neural networks for Al alloy in robotic arc welding," *Journal of Manufacturing Processes,* vol. 45, pp. 208–216, 2019/09/01/2019, https://doi.org/10.1016/j.jmapro.2019.06.023.
23. W. Ren, G. Wen, S. Liu, Z. Yang, B. Xu, and Z. Zhang, "Seam Penetration Recognition for GTAW Using Convolutional Neural Network Based on Time-Frequency Image of Arc Sound," in *2018 IEEE 23rd International Conference on Emerging Technologies and Factory Automation (ETFA),* 4–7 Sept. 2018 2018, vol. 1, pp. 853–860, https://doi.org/10.1109/ETFA.2018.8502478.
24. D. Wu, H. Chen, Y. Huang, and S. Chen, "Online Monitoring and Model-Free Adaptive Control of Weld Penetration in VPPAW Based on Extreme Learning Machine," *IEEE Transactions on Industrial Informatics,* vol. 15, no. 5, pp. 2732–2740, 2019, https://doi.org/10.1109/TII.2018.2870933.
25. D. Zhang, L. Qian, B. Mao, C. Huang, B. Huang, and Y. Si, "A Data-Driven Design for Fault Detection of Wind Turbines Using Random Forests and XGboost," *IEEE Access,* vol. 6, pp. 21020–21031, 2018, https://doi.org/10.1109/ACCESS.2018.2818678.
26. W. Chen *et al.*, "A comparative study of logistic model tree, random forest, and classification and regression tree models for spatial prediction of landslide susceptibility," *CATENA,* vol. 151, pp. 147–160, 2017/04/01/ 2017, https://doi.org/10.1016/j.catena.2016.11.032.
27. T. Chen and C. Guestrin, "Xgboost: A scalable tree boosting system," 2016, pp. 785–794.
28. Y. Zhou, T. Li, J. Shi, and Z. Qian, "A CEEMDAN and XGBOOST-Based Approach to Forecast Crude Oil Prices," *Complexity,* vol. 2019, p. 4392785, 2019/02/03 2019, https://doi.org/10.1155/2019/4392785.
29. S. S. Dhaliwal, A.-A. Nahid, and R. Abbas, "Effective Intrusion Detection System Using XGBoost," *Information*, vol. 9, no. 7, https://doi.org/10.3390/info9070149.
30. X. Zhang, S. T. Waller, and P. Jiang, "An ensemble machine learning-based modeling framework for analysis of traffic crash frequency," *Computer-Aided Civil and Infrastructure Engineering,* vol. 35, no. 3, pp. 258–276, 2020/03/01 2020, https://doi.org/10.1111/mice.12485.

31. S. Gupta, G. Arango-Argoty, L. Zhang, A. Pruden, and P. Vikesland, "Identification of discriminatory antibiotic resistance genes among environmental resistomes using extremely randomized tree algorithm," *Microbiome,* vol. 7, no. 1, p. 123, 2019/08/29 2019, https://doi.org/10.1186/s40168-019-0735-1.
32. J. Park *et al.*, "Reconstruction of Ocean Color Data Using Machine Learning Techniques in Polar Regions: Focusing on Off Cape Hallett, Ross Sea," *Remote Sensing*, vol. 11, no. 11, https://doi.org/10.3390/rs11111366.
33. F. Zhang, J. Bai, X. Li, C. Pei, and V. Havyarimana, "An ensemble cascading extremely randomized trees framework for short-term traffic flow prediction," *KSII Transactions on Internet and Information Systems (TIIS),* vol. 13, no. 4, pp. 1975–1988, 2019.
34. A. Rubio-Solis and G. Panoutsos, "An ensemble data-driven fuzzy network for laser welding quality prediction," in *2017 IEEE International Conference on Fuzzy Systems (FUZZ-IEEE),* 9–12 July 2017 2017, pp. 1–6, https://doi.org/10.1109/FUZZ-IEEE.2017.8015496.
35. L. Breiman, Random forests[J]. Machine learning, 2001, 45:5–32.
36. J. H. Friedman, Stochastic gradient boosting[J]. Computational statistics & data analysis, 2002, 38(4):367–378.

Chapter 5
Summary and Conclusions

5.1 Conclusions

Spectral information and arc sound data provide crucial insights into the material state and energy conversion during welding processes, including plasma temperature, component concentrations, and arc stability. These data are essential for real-time monitoring of welding quality and stability. Additionally, by analyzing this information, it is possible to promptly identify and alert potential defects and anomalies that may occur during welding processes. Subsequently, timely adjustments of welding parameters or corresponding measures can be taken to prevent the occurrence or expansion of defects, thereby improving welding quality and efficiency. Furthermore, utilizing spectral information and arc sound data for intelligent welding monitoring enables automated control of welding processes, reducing manual intervention and enhancing production efficiency and consistency.

The book provides a detailed description of the application of spectral information and arc sound data in intelligent welding monitoring. In summary, the key conclusions of the book can be outlined as follows:

1. The book reveals the arc sound characteristics of different weld penetration states and proposes a convolutional neural network weld penetration state recognition model based on the time-frequency matrix of arc sound. This model achieves accurate recognition of four weld penetration states in aluminum alloy gas tungsten arc welding: unwelded, normal, over-penetration, and burn-through. Initially, the arc sound is divided into pulse and oscillation decay phases, and the frequency energy distribution range of arc sound corresponding to different pulse phases is analyzed separately. The results indicate that the arc energy in the burn-through state leaks from the bottom of the weld pool, resulting in weaker temporal and spectral magnitudes of the arc sound. Furthermore, by using the short-time Fourier transform to convert the time-varying arc sound signal into a time-frequency matrix, combined with CNN feature depth mining and high-dimensional data processing capabilities, the model achieves adaptive mining of

Z. Zhang et al., *Multiple Sensing-Driven Feature Extraction and Defect Detection*,
https://doi.org/10.1007/978-981-96-4808-5_5

deep features from the arc sound time-frequency matrix and weld penetration state recognition. Additionally, the robustness and reliability of the method are validated through feature-by-feature visualization techniques.

2. Online monitoring of internal porosity in gas tungsten arc welding was achieved based on spectral information. This book conducted feature selection and dimensionality reduction on welding spectral information under different porosity conditions. Firstly, FastICA was used for feature extraction of important elemental spectral lines, selecting and optimizing representative spectral features from high-dimensional spectral data. Subsequently, the correlation between important spectral lines and different porosity rates was established, analyzing the variation patterns of different metallic elements. Finally, PCA was employed for dimensionality reduction of high-dimensional features, enabling effective management and analysis of spectral information under different porosity conditions during the welding process. PCA can help transform high-dimensional spectral feature data into a lower-dimensional representation, retaining the most important information while reducing data complexity, thus providing more manageable data for subsequent data processing and model establishment. Finally, ensemble learning learns and represents the distribution of different internal porosity from dimensionality reduction features. Compared to other algorithms, ensemble learning is simple, efficient, reliable, and has excellent feature importance evaluation function. The combination of dimensionality reduction algorithms and ensemble learning algorithms can ensure both recognition accuracy and reliability.
3. This book presents a monitoring framework that achieves the identification of complex welding defects using spectral information. Through research on selected metal spectral bands including line selection, sensitivity evaluation, feature extraction, dynamic response, and process monitoring, the proposed framework exhibits excellent monitoring performance. Firstly, a feature extraction method based on principal component analysis coefficients of interested spectral lines is proposed, analyzing the reliability of sensitivity for different elemental spectral lines. It was found that the spectral features of peak current have better sensitivity to weld defects compared to base current. Subsequently, 50 ratio features of the spectrum were computed, and a quantitative index of feature importance was proposed, contributing to reducing the original feature space and enhancing the robustness of spectral features. Based on key spectral features, model training and optimization were conducted using the random forest algorithm. The use of the random forest algorithm enables the model to have high accuracy and robustness, effectively identifying complex welding defects. Additionally, the monitoring framework also adopts dynamic response and process detection techniques.
4. In order to achieve non-destructive real-time measurement of the weld strength in laser welding of aluminum-lithium alloy, this book proposes a new method based on spectral and hybrid ensemble learning models. Specifically, the book first preliminarily selects different elemental spectra and analyzes the correlation and variable patterns between sensitive spectral features and weld strength. Then,

a spectral data features mining method called RFICA is proposed. This method helps to establish independent feature sets with low correlation and high diversity. Based on this, the average classification accuracy of the RFICA-ET model can be improved by 64.6%. The classification process and mechanism are visualized through decision trees, enhancing the credibility of their decisions. Additionally, the book proposes a hybrid model called RFPCA-XGBoost for predicting laser welding strength. RFPCA exhibits good complementarity, revealing more hidden information about high-dimensional spectral features, while XGBoost demonstrates excellent classification recognition ability. The proposed methods are scalable and can be applied to other welding process sites.

5. This book reveals the plasma composition and periodic fluctuation characteristics in the 7075-aluminum alloy Directed Energy Deposition process. Further analysis indicates that the periodic plasma fluctuations are mainly caused by the absorption and scattering effects of the DED plasma on the laser. A multi-dimensional spectral feature-based intelligent detection method for local pores in aluminum alloy multi-layer DED is proposed, enabling the extraction of plasma shape features under different brightness levels. Based on this, it is found that spectral and plasma image features exhibit good responses to formation defects, and the response characteristics of spectral features to process parameters are analyzed. The mechanisms for the generation of different types of pores due to stochastic and thermal accumulation aggregation are analyzed. The former is caused by random factors such as laser energy fluctuations and spatter during the additive manufacturing process, resulting in an intense processing state and a significant enhancement of both background and line spectra. As the deposited specimens are continuously printed, thermal accumulation leads to an increase in the melt pool temperature and the evaporation of metal elements, reducing the stability of the processing process and consequently increasing the porosity, accompanied by an overall enhancement of spectral line and background intensities. The proposed method exhibits strong recognition capabilities for samples from the same or adjacent deposition layers, ultimately achieving unsupervised recognition of single-layer DED forming and comprehensive pore defects.

5.2 Future Research Directions and Challenges

In the realm of intelligent welding monitoring utilizing spectral and arc sound information, future research directions and challenges encompass several aspects, including:

1. Multimodal Information Fusion: Future research can explore how to better integrate spectral information and arc sound information, along with other potential sensor data such as temperature, pressure, etc., to achieve fusion and joint analysis of multimodal information. This will enhance the comprehensiveness and accuracy of welding process monitoring.

2. Intelligent Algorithm and Model Optimization: Given the large volume and high complexity of data in welding processes, future research can further optimize intelligent algorithms and models, including deep learning, machine learning, and fuzzy logic, to enhance the performance and stability of monitoring systems. This will enable more accurate defect detection and prediction.
3. Real-time Monitoring and Feedback Control: Future welding intelligent monitoring systems can pay more attention to real-time capabilities and feedback control. By collecting and processing data in real-time, anomalies in the welding process can be detected promptly, and adjustments can be made dynamically to control the welding quality and stability.
4. Data Security and Privacy Protection: With the increase in data volume involved in welding monitoring systems, data security and privacy protection will become important challenges. Future research needs to focus on how to ensure the security and privacy of welding process data while ensuring the reliability and stability of monitoring systems.
5. Standardization and Regulation: To promote and apply welding intelligent monitoring technology, future research needs to strengthen standardization and regulation efforts. Establishing unified monitoring methods and evaluation standards is essential to ensure the reliability and comparability of monitoring systems and promote their widespread application in industrial practice.
6. Material and Process Optimization: Based on welding monitoring data and intelligent analysis, future research can explore how to optimize welding materials and process parameters to improve welding quality and efficiency, and reduce energy consumption and material waste.
7. Remote Monitoring and Control: With the development of the Internet of Things and remote communication technology, welding intelligent monitoring systems can achieve remote monitoring and control. This enables real-time monitoring and control of the welding process at remote locations, thereby achieving automation and intelligence in the production process.
8. Intelligent Fault Diagnosis System: Based on big data and machine learning technology, establish an intelligent fault diagnosis system that can automatically identify various abnormal situations in the welding process and provide targeted fault diagnosis and repair recommendations.
9. Application in New Materials and Processes: With the emergence of new materials and welding processes, future research can apply welding intelligent monitoring technology to the study of new materials and processes to address the challenges and opportunities brought by new materials and processes.
10. Sustainable Development and Environmental Protection: Welding intelligent monitoring technology can also be combined with sustainable development and environmental protection. By optimizing welding processes and material selection, the impact of welding processes on the environment can be reduced, and resource utilization efficiency and energy utilization efficiency can be improved.

11. Adaptive Optimization Algorithms: Research and develop optimization algorithms that can adaptively adjust parameters and strategies to cope with changes in materials, thicknesses, and environmental conditions during the welding process, achieving optimal welding quality and efficiency.

MIX
Papier aus verantwortungsvollen Quellen
Paper from responsible sources
FSC® C105338
FSC
www.fsc.org

If you have any concerns about our products,
you can contact us on
ProductSafety@springernature.com

In case Publisher is established outside the EU,
the EU authorized representative is:
Springer Nature Customer Service Center GmbH
Europaplatz 3, 69115 Heidelberg, Germany

Printed by Libri Plureos GmbH
in Hamburg, Germany